Lecture Notes
in Control and Information Sciences 347

Editors: M. Thoma, M. Morari

Wudhichai Assawinchaichote,
Sing Kiong Nguang, Peng Shi

Fuzzy Control and Filter Design for Uncertain Fuzzy Systems

Springer

Authors

Wudhichai Assawinchaichote
Department of Electronic and
Telecommucation Engineering
King Mongkut's University of
Technology Thonburi
91 Prachautits Rd., 10140
Bangkok, Thailand
E-mail: wudhichai.asa@kmutt.ac.th

Peng Shi
Faculty Technology Studies
Department of Mathematics
University Glamorgan
CF37 1DL Pontypridd
M. Glam., United Kingdom
E-mail: pshi@glam.ac.uk

Sing Kiong Nguang
Department of Electronical and
Computer Engineering
The University of Auckland
Private Bag 92019
Auckland, New Zealand
E-mail: sk.nguang@auckland.ac.nz

Library of Congress Control Number: 2006930672

ISSN print edition: 0170-8643
ISSN electronic edition: 1610-7411
ISBN-10 3-540-37011-0 Springer Berlin Heidelberg New York
ISBN-13 978-3-540-37011-6 Springer Berlin Heidelberg New York

Springer is a part of Springer Science+Business Media
springer.com

Typesetting: by the authors and techbooks using a Springer LaTeX macro package

Cover design: *design & production* GmbH, Heidelberg

Printed on acid-free paper SPIN: 11777625 89/techbooks 5 4 3 2 1 0

Preface

Most real physical systems are nonlinear in nature. Control and filtering of nonlinear systems are still open problems due to their complexity natures. These problem becomes more complex when the system's parameters are uncertain. A common approach to designing a controller/filter for an uncertain nonlinear system is to linearize the system about an operating point, and uses linear control theory to design a controller/filter. This approach is successful when the operating point of the system is restricted to a certain region. However, when a wide range operation of the system is required, this method may fail.

This book presents new novel methodologies for designing robust $\mathcal{H}_\infty$ fuzzy controllers and robust $\mathcal{H}_\infty$ fuzzy filters for a class of uncertain fuzzy systems (UFSs), uncertain fuzzy Markovian jump systems (UFMJSs), uncertain fuzzy singularly perturbed systems (UFSPSs) and uncertain fuzzy singularly perturbed systems with Markovian jumps (UFSPS–MJs). These new methodologies provide a framework for designing robust $\mathcal{H}_\infty$ fuzzy controllers and robust $\mathcal{H}_\infty$ fuzzy filters for these classes of systems based on a Tagaki-Sugeno (TS) fuzzy model. Solutions to the design problems are presented in terms of linear matrix inequalities (LMIs). To investigate the design problems, we first describe a class of uncertain nonlinear systems (UNSs), uncertain nonlinear Markovian jump systems (UNMJSs), uncertain nonlinear singularly perturbed systems (UNSPSs) and uncertain nonlinear singularly perturbed systems with Markovian jumps (UNSPS–MJs) by a TS fuzzy system with parametric uncertainties and with/without Markovian jumps. Then, based on an LMI approach, we develop a technique for designing robust $\mathcal{H}_\infty$ fuzzy controllers and robust $\mathcal{H}_\infty$ fuzzy filters such that a given prescribed performance index is guaranteed.

To clarify this approach, this book is divided into two parts. Part I is focused on the uncertain fuzzy systems, while Part II is concentrated on the uncertain fuzzy singularly perturbed systems. Contributions of each part can be summarized as follows. Part I presents the new design methodology on a robust $\mathcal{H}_\infty$ fuzzy controller and a robust $\mathcal{H}_\infty$ fuzzy filter for a class of UFSs and UFMJSs. This new design approach allows us to achieve the design of the robust $\mathcal{H}_\infty$ fuzzy controller and fuzzy filter for a class of UNSs and UNMJSs in a unified framework, which is based on the TS fuzzy model and the LMIs approach. The proposed design result satisfies all admissible noises, disturbances and uncertainties. In parallel, Part II presents the design methodology on a robust $\mathcal{H}_\infty$ fuzzy controller and a robust $\mathcal{H}_\infty$ fuzzy filter for a class of UFSPSs and UFSPS–MJs. The proposed design approach in this part for a class of UFSPSs and UFSPS–MJs does not involve the separation of states into slow and fast ones and it can be applied not only to standard, but also to nonstandard nonlinear singularly perturbed systems. Furthermore, the proposed approach shows that the design result satisfies all admissible noises, disturbances and uncertainties.

Finally, to demonstrate the effectiveness and advantages of the proposed design methodologies, applications to UNS, UNMJS, UNSPS and UNSPS–MJ (for instance; a motor, a tunnel diode circuit and an economic model) are given as examples in each chapter. The simulation results show that the proposed design methodologies can achieve the prescribed performance index.

List of Abbreviations

ARE	Algebraic Riccati Equation
FLC	Fuzzy Logic Control
HJE	Hamilton-Jacobi Equation
LMI	Linear Matrix Inequality
LMJS	Linear Markovian Jump System
LSPS	Linear Singularly Perturbed System
LSPS–MJ	Linear Singularly Perturbed System with Markovian Jump
MJ	Markovian Jump
MJS	Markovian Jump System
NMJS	Nonlinear Markovian Jump System
NS	Nonlinear System
NSPS	Nonlinear Singularly Perturbed System
NSPS–MJ	Nonlinear Singularly Perturbed System with Markovian Jump
SPS	Singularly Perturbed System
SPS–MJ	Singularly Perturbed System with Markovian Jump
TS	Takagi-Sugeno
UFS	Uncertain Fuzzy System
UFMJS	Uncertain Fuzzy Markovian Jump System
UFSPS	Uncertain Fuzzy Singularly Perturbed System
UFSPS–MJ	Uncertain Fuzzy Singularly Perturbed System with Markovian Jump
UNS	Uncertain Nonlinear System
UNMJS	Uncertain Nonlinear Markovian Jump System
UNSPS	Uncertain Nonlinear Singularly Perturbed System
UNSPS–MJ	Uncertain Nonlinear Singularly Perturbed System with Markovian Jump

Contents

1 Introduction 1
1.1 Preliminary Background 1
1.2 Motivation 3
1.3 Contribution of the Book 4
1.4 Book Organization 5

Part I UNCERTAIN FUZZY SYSTEMS

2 Uncertain Fuzzy Systems 9
2.1 Background and Motivation 9
2.1.1 TS Fuzzy Modelling 10
2.1.2 TS fuzzy Controller 14
2.2 Outline of Part I 16

3 Robust $\mathcal{H}_\infty$ Fuzzy Control Design for Uncertain Fuzzy Systems 17
3.1 System Description 17
3.2 Robust $\mathcal{H}_\infty$ State-Feedback Control Design 18
3.3 Robust $\mathcal{H}_\infty$ Output Feedback Control Design 19
3.3.1 Case I–$\nu(t)$ is available for feedback 20
3.3.2 Case II–$\nu(t)$ is unavailable for feedback 23
3.4 Example 26
3.5 Conclusion 31

4 Robust $\mathcal{H}_\infty$ Fuzzy Filter Design for Uncertain Fuzzy Systems 33
4.1 Robust $\mathcal{H}_\infty$ Fuzzy Filter Design 33
4.1.1 Case I–$\nu(t)$ is available for feedback 34
4.1.2 Case II–$\nu(t)$ is unavailable for feedback 38
4.2 Example 40

4.3 Conclusion 45

5 **Robust $\mathcal{H}_\infty$ Fuzzy Control Design for Uncertain Fuzzy Markovian Jump Systems** 47
5.1 System Description 47
5.2 Robust $\mathcal{H}_\infty$ State-Feedback Control Design 49
5.3 Robust $\mathcal{H}_\infty$ Output Feedback Control Design 50
5.3.1 Case I–$\nu(t)$ is available for feedback 50
5.3.2 Case II–$\nu(t)$ is unavailable for feedback 54
5.4 Example 57
5.5 Conclusion 64

6 **Robust $\mathcal{H}_\infty$ Fuzzy Filter Design for Uncertain Fuzzy Markovian Jump Systems** 65
6.1 Robust $\mathcal{H}_\infty$ Fuzzy Filter Design 65
6.1.1 Case I–$\nu(t)$ is available for feedback 66
6.1.2 Case II–$\nu(t)$ is unavailable for feedback 69
6.2 Example 71
6.3 Conclusion 78

Part II UNCERTAIN FUZZY SINGULARLY PERTURBED SYSTEMS

7 **Uncertain Fuzzy Singularly Perturbed Systems** 81
7.1 Background and Motivation 81
7.2 Outline of Part II 83

8 **Robust $\mathcal{H}_\infty$ Fuzzy Control Design for Uncertain Fuzzy Singularly Perturbed Systems** 85
8.1 System Description 85
8.2 Robust $\mathcal{H}_\infty$ State-Feedback Control Design 86
8.3 Robust $\mathcal{H}_\infty$ Output Feedback Control Design 89
8.3.1 Case I–$\nu(t)$ is available for feedback 89
8.3.2 Case II–$\nu(t)$ is unavailable for feedback 92
8.4 Example 94
8.5 Conclusion 99

9 **Robust $\mathcal{H}_\infty$ Fuzzy Filter Design for Uncertain Fuzzy Singularly Perturbed Systems** 101
9.1 Robust $\mathcal{H}_\infty$ Fuzzy Filter Design 101
9.1.1 Case I–$\nu(t)$ is available for feedback 102
9.1.2 Case II–$\nu(t)$ is unavailable for feedback 104
9.2 Example 106
9.3 Conclusion 111

10 Robust $\mathcal{H}_\infty$ Fuzzy Control Design for Uncertain Fuzzy Singularly Perturbed Systems with Markovian Jumps 113
10.1 System Description 113
10.2 Robust $\mathcal{H}_\infty$ State-Feedback Control Design 114
10.3 Robust $\mathcal{H}_\infty$ Fuzzy Output Feedback Control Design 118
10.3.1 Case I–$\nu(t)$ is available for feedback 118
10.3.2 Case II–$\nu(t)$ is unavailable for feedback 122
10.4 Example 125
10.5 Conclusion 133

11 Robust $\mathcal{H}_\infty$ Fuzzy Filter Design for Uncertain Fuzzy Singularly Perturbed Systems with Markovian Jumps 135
11.1 Robust $\mathcal{H}_\infty$ Fuzzy Filter Design 135
11.1.1 Case I–$\nu(t)$ is available for feedback 136
11.1.2 Case II–$\nu(t)$ is unavailable for feedback 139
11.2 Example 141
11.3 Conclusion 148

A Proof 149

References 169

1

Introduction

1.1 Preliminary Background

Most real physical systems are nonlinear in nature. Controlling and filtering nonlinear systems are still open problems due to their complexity natures. These problem becomes more complex when the system's parameters are uncertain. A common approach to designing a controller/filter for an system is to linearize the system about an operating point, and uses linear control theory to design a controller/filter. This approach is successful when the operating point of the system is restricted to a certain region. However, when a wide range operation of the system is required, this method may fail.

In recent years, a large number of practical control problems have involved designing a controller/filter that minimizes the worst-case ratio of output energy (filter error energy) to disturbance energy which is known as an $\mathcal{H}_\infty$ control (filter) problem. The $\mathcal{H}_\infty$ control or filter problem is able to address the issue of system parameter uncertainty, and also be applied to the typical problem of disturbance input control. So far, in the literature, the problem of nonlinear $\mathcal{H}_\infty$ control and filter has been extensively studied by a number of researchers; e.g., [1, 2, 3, 4, 5, 6, 7, 8, 9, 10, 11, 12, 13, 14, 15, 16, 17, 18, 19]. In general, there are two commonly used approaches for providing solutions to nonlinear $\mathcal{H}_\infty$ control and filter problems. One is based on the dissipativity theory and theory of differential games; see [1, 4], and the other is based on the nonlinear version of classical Bounded Real Lemma as developed by Willems [20], and Hill and Moylan [21]; also see [6, 7, 22]. Both of these approaches convert the problem of nonlinear $\mathcal{H}_\infty$ control to the solvability of the so-called Hamilton-Jacobi equation (HJE). Recently, the interconnection between the robust nonlinear $\mathcal{H}_\infty$ control (filter) problem and the nonlinear $\mathcal{H}_\infty$ control (filter) problem in terms of a scaled HJE has been studied; e.g., [23, 24, 25, 26]. A good feature of these results is that they are parallel to the linear $\mathcal{H}_\infty$ results. Even, until now, it is very difficult to solve for a global solution to the HJE either analytically or numerically [27, 28]. In general, a globally smooth solution to the HJE cannot be solved because the derivative may experience discontinuity across certain lower-dimensional set [29].

In the last two decades, various schemes have been developed to overcome the aforementioned difficulties in the design of a controller and filter for an uncertain nonlinear system, among which a successful approach is fuzzy logic control (FLC). FLC was first introduced as the foundation of the linguistic model by L.A. Zadeh, [30], in 1965. Zadeh also proposed the fuzzy set theory to provide a tool to solve an ill-defined problem. Until 1973, Zadeh, [31], outlined the basic concept underlying FLC which is the linguistic variable, the fuzzy IF-THEN rules, the fuzzy algorithm, the composition rule of inference, and the execution of fuzzy instructions. FLC has been considered as an efficient and effective tool in managing uncertainties and nonlinearity of the system since Zadeh's seminar paper, [30]. Among many applications of FLC appear to be one that has attracted a large number of researchers in the past two decades with many successful applications. The work of Mamdani and Asilian, [32], in 1975 showed the first practical application of FLC that implemented Zadeh's fuzzy set theory. This method is known as the Mamdani model. The other method is the Takagi-Sugeno (TS) fuzzy model which was introduced by Takagi and Sugeno, [33], in 1985.

The Mamdani type fuzzy model is often called a pure fuzzy model and consists of a series of rules that cover most of the state space of the system. Each rule maps fuzzy inputs to fuzzy outputs. While the TS fuzzy model also consists of fuzzy rules, the input fuzzy rules are used to select functions of the input variables for outputs. In this way, the mathematical model of the system can be more directly implemented in the fuzzy model. The TS fuzzy model typically acts to select various linear equations over the state space of the system and provides a smooth transition between each one. Generally speaking, the TS fuzzy model is a nonlinear model consisting of a number of rule-based linear models and membership functions which determine the degrees of confidence of the rule. The TS fuzzy model can be used to approximate global behavior of a highly complex nonlinear system. In the TS fuzzy model, local dynamics in different state space regions are represented by local linear systems. The overall model of the system is obtained by "blending" these linear models through nonlinear fuzzy membership functions.

Recently, there have been a number of researchers studying the TS fuzzy system and control; e.g., [34, 35, 27, 36, 37, 38, 39, 40, 41, 42, 43, 44, 45, 46, 47, 48, 49, 50, 51, 52, 53]. For example, Tanaka et.al., [43], have investigated the fuzzy control based on quadratic performance function–an LMI approach, and Nguang and Shi, [27], have considered the $\mathcal{H}_\infty$ fuzzy output feedback control design for nonlinear systems based on an LMI approach. Fuzzy control systems have proven to be superior in performance when compared with conventional control systems especially in controlling nonlinear, ill-defined system and in managing complex system.

1.2 Motivation

Key motivation to this book comes from several sources. The most general motivation is that most real physical systems are nonlinear and parametric uncertainties often exist in the systems, but until now we do not have a systematic way for designing a robust $\mathcal{H}_\infty$ fuzzy controller and filter for it. So far, although the HJE–based sufficient conditions for nonlinear systems to have an $\mathcal{H}_\infty$ performance has been derived, it is still very difficult to find a global solution to the HJE either analytically or numerically; e.g., [27, 28].

A second motivation arises from the successful approach of FLC that overcomes the design problem for nonlinear systems. Fuzzy system theory enables us to utilize qualitative, linguistic information about a highly complex nonlinear system to construct a mathematical model for it. Recent studies show that a TS fuzzy model can be used to approximate global behaviors of a highly complex nonlinear system; e.g., [34]–[53]. In this thesis, the TS fuzzy model has been chosen in our investigation due to the fact that the TS fuzzy model can replace the fuzzy sets in the consequent part of the Mamdani rule with a series of linear equations of input variables with smooth transition between each one. Recently, a great amount of effort has been made on the design of fuzzy $\mathcal{H}_\infty$ for a class of NSs which can be represented by the TS fuzzy model; e.g., [30, 31, 34, 35, 27, 38, 39, 40, 41, 54, 55, 56, 57, 58]. In the TS fuzzy model, local dynamics in different state space regions are represented by local linear systems. The overall model of the system is obtained by "blending" these linear models through nonlinear fuzzy membership functions. Unlike conventional modelling which uses a single model to describe the global behavior of a system, fuzzy modelling is essentially a multi-model approach in which simple sub-models (linear models) are combined to describe the global behavior of the system.

However, it is also necessary for us to further consider the robust stability against parametric uncertainties in the TS fuzzy model. This is because the parametric uncertainties play an important factor responsible for the stability and performance of an uncertain nonlinear control system. So far, there have been some attempts in the area of UNSs based on the TS fuzzy model in the literature; e.g., [41, 48, 59]. However, these existing design methods have not distinguished nonlinearity from uncertainty when analyzing the design problems which makes the results conservative. Hence, we need to investigate and find a new technique for designing a robust $\mathcal{H}_\infty$ fuzzy control and filter of a class of UNSs by distinguishing nonlinearity from uncertainty.

The final and somewhat peripheral motivation is that many control design problems are normally formulated in terms of inequalities rather than simple equalities and a lot of problems in control engineering systems can be formulated as LMI feasibility problems; e.g., [60, 61, 62, 63, 64]. Some common convex programming tools, such as ellipsoid methods, interior point methods and methods of alternating convex projections, can be applied to solve the LMIs. However, the interior-point method has been proven that it is extremely efficient in solving the LMI with significant computational complexity. The LMI framework provides a tractable method to solve the problems which has either analytical solution or non-analytical solution. Furthermore, a very powerful and efficient toolbox in MATLAB has been available for solving LMI feasible and optimization problems by interior point methods.

1.3 Contribution of the Book

The focus of this book is to establish novel methodologies for designing robust $\mathcal{H}_\infty$ fuzzy controllers and robust $\mathcal{H}_\infty$ fuzzy filters for a class of UNSs, UNMJSs, UNSPSs and UNSPS–MJs which are described by a TS fuzzy system with parametric uncertainties and with/without MJs. In this book, we distinguish nonlinearities from uncertainties when analyzing the design problems. This is because if the design problems for a system are analyzed by treating nonlinearities as uncertainties, the results will be conservative in general; e.g., [48, 59]. The derivations of the solutions shown here are based on fuzzy control theory and robust control theory such as an $\mathcal{H}_\infty$ control and an LMI.

To investigate the design problems, we first describe a class of UNSs, UNMJSs, UNSPSs and UNSPS–MJs by a TS fuzzy system with parametric uncertainties and with/without MJs. Then, based on an LMI approach, we develop a technique for designing robust $\mathcal{H}_\infty$ fuzzy controllers and robust $\mathcal{H}_\infty$ fuzzy filters such that a given prescribed performance index is guaranteed. To understand our proposed approach better, this thesis has been divided into two parts. Part I is focused on the uncertain fuzzy systems, while Part II is concentrated on the uncertain fuzzy singularly perturbed systems. The achievement in each part can be summarized as follows:

In Part I, the new design methodology on a robust $\mathcal{H}_\infty$ fuzzy controller and a robust $\mathcal{H}_\infty$ fuzzy filter for a class of UFSs and UFMJSs has been developed. This new design approach allows us to achieve designing the robust $\mathcal{H}_\infty$ fuzzy controller and the robust $\mathcal{H}_\infty$ fuzzy filter for a class of UNSs and UNMJSs in a unified framework, which is based on the TS fuzzy model and LMIs approach. The proposed design result satisfies all admissible noises, disturbances and uncertainties.

In Part II, the design methodology on a robust $\mathcal{H}_\infty$ fuzzy controller and a robust $\mathcal{H}_\infty$ fuzzy filter for a class of UFSPSs and UFSPS–MJs has been presented. The proposed design approach in this part for a class of UFSPSs and UFSPS–MJs does not involve the separation of states into slow and fast ones and it can be applied not only to standard, but also to nonstandard nonlinear singularly perturbed systems. Furthermore, the proposed approach shows that the design result satisfies all admissible noises, disturbances and uncertainties.

Finally, to demonstrate the effectiveness and advantages of the proposed design methodologies in this thesis, applications to UNS, UNMJS, UNSPS and UNSPS–MJ (i.e., a motor, a tunnel diode circuit and an economic model) are given as examples in each chapter. The simulation results show that the proposed design methodologies can achieve the prescribed performance index.

1.4 Book Organization

The general layout of presentation of this book is divided into two parts; i.e., Part I: Uncertain Fuzzy Systems and Part II: Uncertain Fuzzy Singularly Perturbed Systems.

Part I presents the synthesis design procedure of a robust $\mathcal{H}_\infty$ fuzzy control and filter for a class of UFSs and UFMJSs. Part I which begins with Chapter 2 consists of five chapters as follows.

Chapter 2 presents some background and motivation in the area of UFSs, and then provides the outline of Part I before proceeding to the main results from Chapter 3 to Chapter 6.

Chapter 3 presents the synthesis design procedure of a robust $\mathcal{H}_\infty$ fuzzy state-feedback control and output feedback control for a class of UFSs. The resulting fuzzy controller shows that it guarantees the $\mathcal{L}_2$-gain of the mapping from the exogenous input noise to the regulated output to be less than the prescribed value.

Chapter 4 presents the synthesis design procedure of a robust $\mathcal{H}_\infty$ fuzzy filter for a class of UFSs such that it guarantees the $\mathcal{L}_2$-gain of the mapping from the exogenous input noise to the estimated error output to be less than the prescribed value.

Chapter 5 presents the synthesis design procedure of a robust $\mathcal{H}_\infty$ fuzzy state-feedback control and output feedback control for a class of UFMJSs. Due to the fact that many real physical systems may experience abrupt changes in their structure and parameters, this chapter will provide a design technique corresponding to that problem. Solutions to the design problem of the robust $\mathcal{H}_\infty$ fuzzy control have been derived in terms of the LMIs.

Chapter 6 presents the synthesis design procedure of a robust $\mathcal{H}_\infty$ fuzzy filter for a class of UFMJSs. The sufficient conditions to the design problem of the robust $\mathcal{H}_\infty$ fuzzy filter have been derived in terms of the LMIs.

Part II presents the synthesis design procedure of a robust $\mathcal{H}_\infty$ fuzzy control and filter for a class of UFSPSs and UFSPS–MJs. Part II which begins with Chapter 7 also consists of five chapters as follows.

Chapter 7 presents some background and motivation in the area of UFSPSs, and then provides the outline of Part II before proceeding to the main results from Chapter 8 to Chapter 11.

Chapter 8 presents the synthesis design procedure of a robust $\mathcal{H}_\infty$ fuzzy state-feedback control and output feedback control for a class of UFSPSs. In the case of having a small "parasitic" parameter in a general nonlinear system, the design result might end up with a solution to a family of ε-dependent LMIs which normally are ill-conditioned when ε is very small. Then, this chapter will provide a design technique to handle these problems.

Chapter 9 presents the synthesis design procedure of a robust $\mathcal{H}_\infty$ fuzzy filter for a class of UFSPSs. Solutions to the design problem of the robust $\mathcal{H}_\infty$ fuzzy filter have been derived in terms of the LMIs.

Chapter 10 presents the synthesis design procedure of a robust $\mathcal{H}_\infty$ fuzzy state-feedback control and output feedback control for a class of UFSPS–MJs. Since many real physical systems may experience not only a small "parasitic" parameter in the system but also abrupt changes in their structures and parameters, in order to deal with these problems, this chapter will provide a design technique for handling with this case.

Chapter 11 presents the synthesis design procedure of a robust $\mathcal{H}_\infty$ fuzzy filter for a class of UFSPS–MJs. The sufficient conditions to the design problem of the robust $\mathcal{H}_\infty$ fuzzy filter have been derived in terms of the LMIs.

Part I

UNCERTAIN FUZZY SYSTEMS

2

Uncertain Fuzzy Systems

2.1 Background and Motivation

In the last decade, TS fuzzy systems have been studied by many researchers with a number of successful applications. The TS fuzzy system has been shown to be an universal approximator of nonlinear dynamic systems; e.g., [49, 51]. The TS fuzzy system is described by fuzzy IF-THEN rules of the following form:

Plant Rule *i:* IF $\nu_1(t)$ is M_{i1} and $\cdots$ and $\nu_\vartheta(t)$ is $M_{i\vartheta}$ THEN

$$\dot{x}(t) = A_i x(t) + B_i u(t), \quad x(0) = 0 \tag{2.1}$$

where $i = 1, 2, \cdots, r$, $M_{ij}(j = 1, 2, \cdots, \vartheta)$ are fuzzy sets that are characterized by membership functions, $x(t) \in \Re^n$ is the state vector, $u(t) \in \Re^m$ is the input vector, the matrices A_i and B_i are of appropriate dimensions, $\nu_1(t), \cdots, \nu_\vartheta(t)$ are premises variables that may be functions of the state variables, and r is the number of IF-THEN rules.

Given a pair $[x(t), u(t)]$, the final fuzzy system is inferred as follows

$$\dot{x}(t) = \sum_{i=1}^{r} \mu_i(\nu(t))[A_i x(t) + B_i u(t)] \tag{2.2}$$

where

$$\mu_i(\nu(t)) = \frac{\varpi_i(\nu(t))}{\sum_{i=1}^{r} \varpi_i(\nu(t))}) \text{ and } \varpi_i(\nu(t)) = \prod_{k=1}^{\vartheta} M_{ik}(\nu_k(t)).$$

$M_{ik}(\nu_k(t))$ is the grade of membership of $\nu_k(t)$ in M_{ik}. It is assumed that

$$\varpi_i(\nu(t)) \geq 0, \quad i = 1, 2, ..., r; \qquad \sum_{i=1}^{r} \varpi_i(\nu(t)) > 0$$

for all t. Therefore,

$$\mu_i(\nu(t)) \geq 0, \quad i = 1, 2, ..., r; \qquad \sum_{i=1}^{r} \mu_i(\nu(t)) = 1$$

for all t. Figure 2.1 shows the structural diagram of the TS fuzzy system.

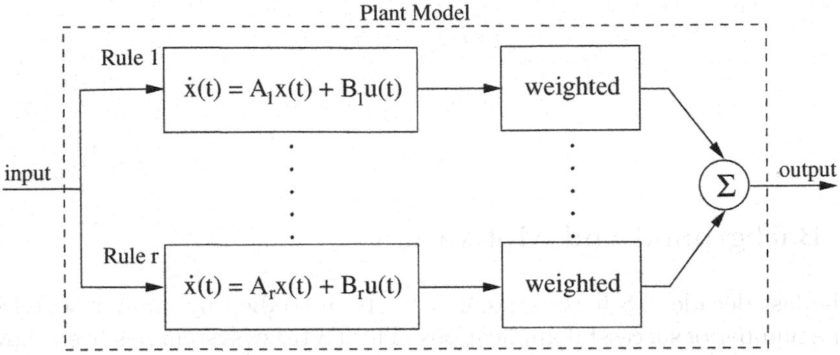

Fig. 2.1. The TS type fuzzy system.

2.1.1 TS Fuzzy Modelling

There are two major ways in TS fuzzy modeling. One is the TS fuzzy model identification ([33], [50] and [65]) using input-output data, and the other is the TS fuzzy model construction, by the idea of sector nonlinearity ([50], [66] and [67]). Fuzzy Modeling and Identification (FMID) toolbox from Matlab can be utilised for the construction of a TS fuzzy model from data. This section only discusses the construction of a TS fuzzy model by the idea of sector nonlinearity. Three examples will be used to illustrate the procedures of constructing TS fuzzy models.

Consider the following class of nonlinear system

$$\dot{x}_i(t) = \sum_{j=1}^{n} f_{ij}(x(t))x_j(t) + \sum_{k=1}^{m} g_{ik}(x(t))u_k(t) \tag{2.3}$$

where n and m are, respectively, the numbers of state variables and inputs. $x(t) = [x_1(t) \quad \cdots \quad x_n(t)]$ is the state vector and $u(t) = [u_1(t) \quad \cdots \quad u_n(t)]$ is the input vector. $f_{ij}(x(t))$ and $g_{ik}(x(t)$ are functions of $x(t)$. To obtained a TS fuzzy model, we find the minimum and maximum values of $f_{ij}(x(t))$ and $g_{ik}(x(t))$,

$$a_{ij1} = \max_{x(t)} \Big\{ f_{ij}(x(t)) \Big\}, \;\; a_{ij2} = \min_{x(t)} \Big\{ f_{ij}(x(t)) \Big\},$$

$$b_{ik1} = \max_{x(t)} \left\{ g_{ik}(x(t)) \right\}, \ b_{ik2} = \min_{x(t)} \left\{ g_{ik}(x(t)) \right\}.$$

By using these variables, f_{ij} and g_{ik} can be represented as

$$f_{ij}(x(t)) = \sum_{\ell=1}^{2} h_{ij\ell}(x(t)) a_{ij\ell}$$

$$g_{ik}(x(t)) = \sum_{\ell=1}^{2} v_{ik\ell}(x(t)) b_{ik\ell}$$

where $\sum_{\ell=1}^{2} h_{ij\ell}(x(t)) = 1$ and $\sum_{\ell=1}^{2} v_{ik\ell}(x(t)) = 1$. The membership functions are assigned as follows:

$$h_{ij1}(x(t)) = \frac{f_{ij}(x(t)) - a_{ij2}}{a_{ij1} - a_{ij2}}, \ h_{ij2}(x(t)) = \frac{a_{ij1} - f_{ij}(x(t))}{a_{ij1} - a_{ij2}}$$

$$g_{ik1}(x(t)) = \frac{g_{ik}(x(t)) - b_{ik2}}{b_{ik1} - b_{ik2}}, \ g_{ik2}(x(t)) = \frac{b_{ij1} - g_{ik}(x(t))}{b_{ik1} - b_{ik2}}$$

By using the fuzzy mode representation, (2.3) can be rewritten as

$$\dot{x}(t) = \sum_{j=1}^{n} \sum_{\ell=1}^{2} h_{ij\ell}(x(t)) a_{ij\ell} x(t) + \sum_{k=1}^{m} \sum_{\ell=1}^{2} v_{ik\ell}(x(t)) b_{ik\ell} u(t). \tag{2.4}$$

Example 1: Lorenz Chaotic System

To design a fuzzy static output feedback controller, the Lorenz chaotic system needs to be represented by a TS fuzzy model. An exact TS fuzzy modelling [66] is employed to construct a TS fuzzy model for the Lorenz chaotic system. The method utilises the concept of sector nonlinearity. For more details, see [50] and [67]. The following Lorenz chaotic system with the input term will be considered in the sequel:

$$\begin{aligned} \dot{x}_1(t) &= -ax_1(t) + ax_2(t) + u(t) \\ \dot{x}_2(t) &= cx_1(t) - x_2(t) - x_1(t)x_3(t) \\ \dot{x}_3(t) &= x_1(t)x_2(t) - bx_3(t) \end{aligned} \tag{2.5}$$

where $a = 10$, $b = 8/3$, $c = 28$, $x_1(t), x_2(t)$ and x_3 are the state variables, and $u(t)$ is the control input. Assume that $x_1(t) \in [-N \ N]$, the nonlinear terms $-x_1(t)x_3(t)$ and $x_1(t)x_2(t)$ can be expressed as

$$-x_1 x_3(t) = -h_1(x_1(t))\big[-Nx_3(t)\big] - h_2(x_1(t))\big[Nx_3(t)\big]$$

and

$$x_1 x_2(t) = h_1(x_1(t))\big[-Nx_2(t)\big] + h_2(x_1(t))\big[Nx_2(t)\big]$$

where

$$h_1(x_1(t)) = \frac{-x_1(t)+N}{2N} \text{ and } h_2(x_1(t)) = \frac{x_1(t)+N}{2N}.$$

Then, using the above membership functions, we can have the following TS fuzzy model which exactly represents (2.5) under the assumption on bounds of the state variable $x_1(t) \in [-N\ N]$.

$$\begin{aligned} \dot{x}(t) &= \textstyle\sum_{i=1}^{2} h_i(x(t))A_i x(t) + \sum_{i=1}^{2} h_i(x(t))B_i u(t) \\ y(t) &= \textstyle\sum_{i=1}^{2} h_i(x(t))C_i x(t) \end{aligned} \tag{2.6}$$

where $x(t) = \left[x_1(t)\ x_2(t)\ x_3(t)\right]^T$,

$$A_1 = \begin{bmatrix} -a & a & 0 \\ c & -1 & -N \\ 0 & N & -b \end{bmatrix}, A_2 = \begin{bmatrix} -a & a & 0 \\ c & -1 & N \\ 0 & -N & -b \end{bmatrix}, B_1 = B_2 = \begin{bmatrix} 1 \\ 0 \\ 0 \end{bmatrix}, C_1 = C_2 \left[1\ 0\ 0\right].$$

The TS fuzzy model (2.6) exactly represents (2.5) under the assumption on bounds of the state variable $x_1(t) \in [-N\ N]$ where $N > 0$. However, this assumption is not strict because of two reasons. It is well know that state variables of the chaotic system are bounded. In addition, N can be set to any value. Even if the nonlinear equations of the Lorenz chaotic system are unknown, recently developed fuzzy modeling techniques ([33], [51] and [65]) using observed data can be utilised to obtain fuzzy models.

Example 2: Nonlinear Mass-Spring-Damper System

Consider a nonlinear mass-spring-damper mechanical system with a nonlinear spring:

$$\begin{aligned} \dot{x}_1(t) &= -0.1125x_1(t) - 0.02x_2(t) - 0.67x_2^3(t) + u(t) \\ \dot{x}_2(t) &= x_1(t) \end{aligned} \tag{2.7}$$

where $x_2(t)$ is the spring's displacement and $x_1(t) = \dot{x}_2(t)$. The term $-0.67x_2^3$ is due to the nonlinearity of the spring. The spring is attached to a fixed wall, therefore the spring's displacement $x_2(t)$ is physically constrained by the length of the spring and the wall. The length of the spring could be any value, in this paper, we assume $x_2(t) \in [-1,\ 1.5]$. The lower limit is the minimum length that the spring can be compressed. Same as Example 1, the concept of sector nonlinearity [66] is employed to construct an exact TS fuzzy model for the mass-spring-damper system. Using the fact that $x_2(t) \in [-1,\ 1.5]$, this nonlinear term can be expressed as

$$-0.67x_2^3(t) = -h_1(x_2(t))\left[0x_2(t)\right] - h_2(x_2(t))\left[1.5075\right]x_2(t)$$

where $h_1(x_2(t)) = 1 - \frac{x^2(t)}{2.25}$ and $h_2(x_2(t)) = \frac{x^2(t)}{2.25}$.

Using $h_1(x_2(t))$ and $h_2(x_2(t))$, we obtain the following TS fuzzy model which exactly represents (2.7) under the assumption on bounds of the state variable $x_2(t) \in [-1\ 1.5]$:

$$\dot{x}(t) = \sum_{i=1}^{2} h_i(x(t))A_i x(t) + \sum_{i=1}^{2} h_i(x(t))B_i u(t) \tag{2.8}$$

where $x(t) = \begin{bmatrix} x_1(t) & x_2(t) \end{bmatrix}^T$,

$$A_1 = \begin{bmatrix} -0.1125 & -0.02 \\ 1 & 0 \end{bmatrix}, A_2 = \begin{bmatrix} -0.1125 & -1.5275 \\ 1 & 0 \end{bmatrix}, B_1 = B_2 = \begin{bmatrix} 1 \\ 0 \end{bmatrix}.$$

Example 3: Three Tanks System

Consider an interconnected tank system which consists of three tanks [68] denoted as Tank I, Tank II, and Tank III. Liquid levels $x_i(t)$ are regulated by manipulating the inlet flow rates. Pumps P_1 and P_2 are used to pump liquid into Tank I and II with flow rates $q_{11}(t)$ and $q_{22}(t)$ that are proportional to control inputs $u_1(t)$ and $u_2(t)$, respectively, as follows

$$q_{ii}(t) = p_{ii}u_i(t), \quad i \in \{1,2\} \tag{2.9}$$

where p_{ii} are the section of the opening valves. Liquids from Tanks I and II flow into Tank III with flow rates $q_1(t)$ and $q_2(t)$, respectively. Liquid flows out from Tank III is at the flow rate, $q_3(t)$. The amount of liquid flowing off by an outlet valve is according to Torricelli's law is

$$q_i(t) = p_i\sqrt{x_i(t)}, \quad i \in \{1,2,3\} \tag{2.10}$$

where $p_i = \rho S_i\sqrt{2g}$, S_i are the section of valves, g is the earth's gravity, and ρ is the liquid's density.

The differential equations that describe this three tank systems are as follows:

$$\begin{aligned}
\dot{x}_1(t) &= -\frac{p_1}{\sqrt{x_1(t)}}x_1(t) + p_{11}u_1(t) \\
\dot{x}_2(t) &= -\frac{p_2}{\sqrt{x_2(t)}}x_2(t) + p_{11}u_1(t) \\
\dot{x}_3(t) &= \frac{p_1}{\sqrt{x_1(t)}}x_1(t) + \frac{p_2}{\sqrt{x_2(t)}}x_1(t) - \frac{p_3}{\sqrt{x_3(t)}}x_3(t)
\end{aligned} \tag{2.11}$$

and the two outputs of the system are

$$\begin{aligned}
y_1(t) = q_3(t) &= \frac{p_3}{\sqrt{x_3(t)}}x_3(t) \\
y_2(t) &= \frac{p_1}{\sqrt{x_1(t)}}x_1(t) - \frac{\alpha p_2}{\sqrt{x_2(t)}}x_2(t)
\end{aligned} \tag{2.12}$$

where the parameter α is assumed to be greater than zero to guarantee a fixed relative concentration. Suppose that, in this example, the nonlinear term

$$f_i(x(t)) = \frac{1}{\sqrt{x_i(t)}} \in [a_1 \quad a_2] \tag{2.13}$$

and let the weighting function

$$w_i(x(t)) = \frac{f_i(x(t)) - a_1}{a_2 - a_1}. \tag{2.14}$$

Thus, the normalized time varying weighting function for each rule can be expressed as

$$\begin{aligned}
h_1(x(t)) &= w_1(x(t))w_2(x(t))w_3(x(t))\\
h_2(x(t)) &= w_1(x(t))w_2(x(t))(1-w_3(x(t)))\\
h_3(x(t)) &= w_1(x(t))(1-w_2(x(t)))w_3(x(t))\\
h_4(x(t)) &= (1-w_1(x(t)))w_2(x(t))w_3(x(t))\\
h_5(x(t)) &= w_1(x(t))(1-w_2(x(t)))(1-w_3(x(t)))\\
h_6(x(t)) &= (1-w_1(x(t)))w_2(x(t))(1-w_3(x(t)))\\
h_7(x(t)) &= (1-w_1(x(t)))(1-w_2(x(t)))w_3(x(t))\\
h_8(x(t)) &= (1-w_1(x(t)))(1-w_2(x(t)))(1-w_3(x(t))).
\end{aligned} \tag{2.15}$$

Using the rule above, we have the following TS fuzzy model under the assumption on bounds of $f_i(x(t)) \in [a_1 \ \ a_2]$

$$\begin{aligned}
\dot{x}(t) &= \textstyle\sum_{i=1}^{8} h_i(x(t))A_i x(t) + \sum_{i=1}^{8} h_i(x(t))B_i u(t)\\
y(t) &= \textstyle\sum_{i=1}^{8} h_i(x(t))C_i x(t)
\end{aligned} \tag{2.16}$$

where $x(t) = \begin{bmatrix} x_1(t) & x_2(t) & x_3(t)\end{bmatrix}^T$, $u(t) = \begin{bmatrix} u_1(t) & u_2(t)\end{bmatrix}^T$, $y(t) = \begin{bmatrix} y_1(t) \ y_2(t)\end{bmatrix}^T$, with for example

$$A_1 = \begin{bmatrix} -p_1a_2 & 0 & 0\\ 0 & -p_2a_2 & 0\\ p_1a_2 & p_2a_2 & -p_3a_2 \end{bmatrix}, B_1 = \begin{bmatrix} p_{11} & 0\\ 0 & p_{22}\\ 0 & 0\end{bmatrix}, C_1 = \begin{bmatrix} 0 & 0 & p_3a_2\\ p_1a_2 & -\alpha p_2a_2 & 0\end{bmatrix},$$

$$A_2 = \begin{bmatrix} -p_1a_2 & 0 & 0\\ 0 & -p_2a_2 & 0\\ p_1a_2 & p_2a_2 & -p_3a_1 \end{bmatrix}, B_2 = \begin{bmatrix} p_{11} & 0\\ 0 & p_{22}\\ 0 & 0\end{bmatrix}, C_2 = \begin{bmatrix} 0 & 0 & p_3a_1\\ p_1a_2 & -\alpha p_2a_2 & 0\end{bmatrix},$$

and, so on.

2.1.2 TS fuzzy Controller

For a fuzzy controller design, it is supposed that the fuzzy system is locally controllable. Then, the local state feedback controller is designed as follows:

Controller Rule *i:* IF $\nu_1(t)$ is M_{i1} and $\cdots$ and $\nu_\vartheta(t)$ is $M_{i\vartheta}$ THEN

$$u(t) = -K_i x(t), \quad \text{for } i = 1, 2, \cdots, r \tag{2.17}$$

where K_i is the controller gain. Then, the final TS fuzzy controller is

$$u(t) = -\sum_{i=1}^{r} \mu_i(\nu(t))K_i x(t). \tag{2.18}$$

The block diagram of the TS fuzzy controller is given in Figure 2.2.

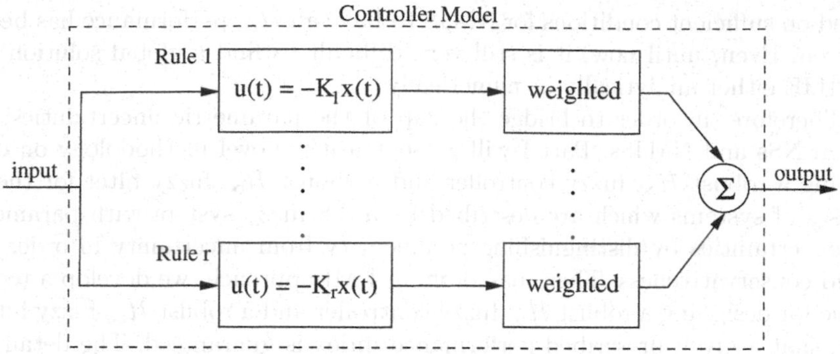

Fig. 2.2. The TS type fuzzy controller.

The resulting fuzzy controller (2.18) is nonlinear in general since the coefficients of the controller depend nonlinearly on the system input and output fuzzy weights. Moreover, the resulting fuzzy controller (2.18) could be represented as a particular form of a gain scheduled controller where the gains are varied as a function of operating conditions. The TS type fuzzy control scheme has a major advantage over the existing crisp gain scheduling scheme. That is, it provides a general method for the interpolation of available local control law into an overall gain scheduling control law which is computationally efficient.

Recently, there have been some attempts for designing a fuzzy controller and a fuzzy filter for a class of uncertain nonlinear systems which is described by a TS fuzzy model with parametric uncertainties; e.g., [41, 48, 59]. This is due to the fact that uncertainties are often a source of instability. However, the existing design results of the TS fuzzy model with parametric uncertainties in [41, 48, 59] are quite conservative since they treat nonlinearities as uncertainties when analyzing the problems. Thus, it is necessary to have an approach that can help us to overcome the conservativeness.

It is also clear that many practical application systems may experience abrupt changes in their structure and parameters, caused by phenomena such as parameters shifting, tracking, and the time required to measure some of the variables at different stages. Such a system can be modelled by a hybrid system with two components in the state vector. The first one which varies continuously is referred to as the continuous state of the system and the second one which varies discretely is referred to as the mode of the system. A special class of hybrid systems known as a Markovian jump system (MJS) has been widely used to model manufacturing systems [69] and communication systems [70]. Although linear Markovian jump systems (LMJSs) have been extensively studied; e.g., [71, 72, 73, 74, 75, 76, 77, 78, 79, 80, 81, 82, 83, 84], to the best of our knowledge, the control design of nonlinear Markovian jump systems (NMJSs) has still not been considered in the literature. Recently, there has been some attempt in this area. In [74], the Hamilton-Jacobi equation (HJE) based on sufficient conditions for NMJS to have an $\mathcal{H}_\infty$ performance has been derived. Even, until now, it is still very difficult to find a global solution to the HJE either analytically or numerically.

Therefore, in order to bridge the gap of the parametric uncertainties issue in NSs and NMJSs, Part I will present a new novel methodology on designing a robust $\mathcal{H}_\infty$ fuzzy controller and a robust $\mathcal{H}_\infty$ fuzzy filter for these classes of systems which are described by a TS fuzzy system with parametric uncertainties by distinguishing nonlinearity from uncertainty in order to avoid conservativeness. Then, based on an LMI approach, we develop a technique for designing a robust $\mathcal{H}_\infty$ fuzzy controller and a robust $\mathcal{H}_\infty$ fuzzy filter such that a given prescribed performance index is guaranteed. The detail of the design problems for UNSs and UNMJSs is presented in Chapter 3 to Chapter 6.

2.2 Outline of Part I

In Part I, the synthesis design procedure of a robust $\mathcal{H}_\infty$ fuzzy controller and a robust $\mathcal{H}_\infty$ filter for a class of UNSs and UNMJSs which is described by a TS fuzzy system with parametric uncertainties and with/without MJs is presented. The outline of Part I is presented as follows. Chapter 2 provides some background and motivation on UFSs. Chapters 3 and 4 present the synthesis design procedure of a robust $\mathcal{H}_\infty$ fuzzy controller and a robust $\mathcal{H}_\infty$ fuzzy filter for the class of UFSs. Then, Chapters 5 and 6 respectively present the synthesis design procedure of a robust $\mathcal{H}_\infty$ fuzzy controller and a robust $\mathcal{H}_\infty$ fuzzy filter for the class of UFMJSs. Finally, to illustrative the effectiveness of the design procedures, a numerical example is also given at the end of each chapter.

3

Robust $\mathcal{H}_\infty$ Fuzzy Control Design for Uncertain Fuzzy Systems

In this chapter, we present a new technique for designing a robust fuzzy state and output feedback controller for a TS fuzzy system with parametric uncertainties. Based on an LMI approach, we develop a technique for designing a robust fuzzy controller such that the $\mathcal{L}_2$-gain of the mapping from the exogenous input noise to the regulated output is less than a prescribed value.

3.1 System Description

In this chapter, we generalize the TS fuzzy system to represent a TS fuzzy system with parametric uncertainties. As in [85], we examine a TS fuzzy system with parametric uncertainties as follows:

$$
\begin{aligned}
\dot{x}(t) &= \textstyle\sum_{i=1}^{r} \mu_i(\nu(t))\Big[[A_i + \Delta A_i]x(t) + [B_{1_i} + \Delta B_{1_i}]w(t) \\
&\quad + [B_{2_i} + \Delta B_{2_i}]u(t)\Big], \qquad x(0) = 0 \\
z(t) &= \textstyle\sum_{i=1}^{r} \mu_i(\nu(t))\Big[[C_{1_i} + \Delta C_{1_i}]x(t) + [D_{12_i} + \Delta D_{12_i}]u(t)\Big] \\
y(t) &= \textstyle\sum_{i=1}^{r} \mu_i(\nu(t))\Big[[C_{2_i} + \Delta C_{2_i}]x(t) + [D_{21_i} + \Delta D_{21_i}]w(t)\Big]
\end{aligned}
\tag{3.1}
$$

where $\nu(t) = [\nu_1(t) \ \cdots \ \nu_\vartheta(t)]$ is the premise variable vector that may depend on states in many cases, $\mu_i(\nu(t))$ denotes the normalized time-varying fuzzy weighting functions for each rule (i.e., $\mu_i(\nu(t)) \geq 0$ and $\sum_{i=1}^{r} \mu_i(\nu(t)) = 1$), ϑ is the number of fuzzy sets, $x(t) \in \Re^n$ is the state vector, $u(t) \in \Re^m$ is the input, $w(t) \in \Re^p$ is the disturbance which belongs to $\mathcal{L}_2[0,\infty)$, $y(t) \in \Re^\ell$ is the measurement, $z(t) \in \Re^s$ is the controlled output, the matrices $A_i, B_{1_i}, B_{2_i}, C_{1_i}, C_{2_i}, D_{12_i}$ and D_{21_i} are of appropriate dimensions, and r is the number of IF-THEN rules. The matrices $\Delta A_i, \Delta B_{1_i}, \Delta B_{2_i}, \Delta C_{1_i}, \Delta C_{2_i}, \Delta D_{12_i}$ and ΔD_{21_i} represent the uncertainties in the system and satisfy the following assumption.

Assumption 3.1

$$\Delta A_i = F(x(t),t)H_{1_i}, \quad \Delta B_{1_i} = F(x(t),t)H_{2_i}, \quad \Delta B_{2_i} = F(x(t),t)H_{3_i},$$
$$\Delta C_{1_i} = F(x(t),t)H_{4_i}, \quad \Delta C_{2_i} = F(x(t),t)H_{5_i}, \quad \Delta D_{12_i} = F(x(t),t)H_{6_i}$$
$$\text{and } \Delta D_{21_i} = F(x(t),t)H_{7_i}$$

where H_{j_i}, $j = 1, 2, \cdots, 7$ are known matrix functions which characterize the structure of the uncertainties. Furthermore, the following inequality holds:

$$\|F(x(t),t)\| \le \rho \tag{3.2}$$

for any known positive constant ρ.

Next, let us recall the following definition.

Definition 1. *Suppose γ is a given positive number. A system (3.1) is said to have an $\mathcal{L}_2$-gain less than or equal to γ if*

$$\int_0^{T_f} z^T(t)z(t)dt \le \gamma^2 \left[\int_0^{T_f} w^T(t)w(t)dt\right], \ x(0) = 0 \tag{3.3}$$

for all $T_f \ge 0$ and $w(t) \in \mathcal{L}_2[0, T_f]$.

3.2 Robust $\mathcal{H}_\infty$ State-Feedback Control Design

The aim of this section is to design a robust $\mathcal{H}_\infty$ fuzzy state-feedback controller of the form

$$u(t) = \sum_{j=1}^{r} \mu_j K_j x(t) \tag{3.4}$$

where K_j is the controller gain, such that the inequality (3.3) holds. The state space form of the fuzzy system model (3.1) with the controller (3.4) is given by

$$\begin{aligned} \dot{x}(t) = \textstyle\sum_{i=1}^r \sum_{j=1}^r \mu_i\mu_j \Big[&[(A_i + B_{2_i}K_j) + (\Delta A_i + \Delta B_{2_i}K_j)]x(t) \\ &+[B_{1_i} + \Delta B_{1_i}]w(t)\Big], \quad x(0) = 0. \end{aligned} \tag{3.5}$$

The following theorem provides sufficient conditions for the existence of a robust $\mathcal{H}_\infty$ fuzzy state-feedback controller. These sufficient conditions can be derived by the Lyapunov approach.

Theorem 1. *Consider the system (3.1). Given a prescribed $\mathcal{H}_\infty$ performance $\gamma > 0$ and a positive constant δ, if there exist a matrix $P = P^T$ and matrices Y_j, $j = 1, 2, \cdots, r$, satisfying the following linear matrix inequalities:*

$$P > 0 \tag{3.6}$$

$$\Omega_{ii} < 0, \quad i = 1, 2, \cdots, r \tag{3.7}$$

$$\Omega_{ij} + \Omega_{ji} < 0, \quad i < j \le r \tag{3.8}$$

where

$$\Omega_{ij} = \begin{pmatrix} A_i P + P A_i^T + B_{2_i} Y_j + Y_j^T B_{2_i}^T & (*)^T & (*)^T \\ \tilde{B}_{1_i}^T & -\gamma I & (*)^T \\ \tilde{C}_{1_i} P + \tilde{D}_{12_i} Y_j & 0 & -\gamma I \end{pmatrix} \tag{3.9}$$

with

$$\tilde{B}_{1_i} = \left[\delta I \; I \; \delta I \; B_{1_i}\right], \quad \tilde{C}_{1_i} = \left[\tfrac{\gamma\rho}{\delta} H_{1_i}^T \; 0 \; \sqrt{2}\lambda\rho H_{4_i}^T \; \sqrt{2}\lambda C_{1_i}^T\right]^T,$$

$$\tilde{D}_{12_i} = \left[0 \; \tfrac{\gamma\rho}{\delta} H_{3_i}^T \; \sqrt{2}\lambda\rho H_{6_i}^T \; \sqrt{2}\lambda D_{12_i}^T\right]^T, \lambda = \left(1 + \rho^2 \sum_{i=1}^{r}\sum_{j=1}^{r}\left[\|H_{2_i}^T H_{2_j}\|\right]\right)^{\frac{1}{2}}$$

then the inequality (3.3) holds. Furthermore, a suitable choice of the fuzzy controller is

$$u(t) = \sum_{j=1}^{r} \mu_j K_j x(t) \tag{3.10}$$

where

$$K_j = Y_j P^{-1}. \tag{3.11}$$

Proof: See Appendix. ■

3.3 Robust $\mathcal{H}_\infty$ Output Feedback Control Design

The nature of the information of the state available to the controller has a major effect on the complexity of the designing problem and of the resulting controller. The state-feedback control design problem is an easier problem in which all information are available. However, in most real physical systems, the state is not perfectly known, and so we must estimate it. The process of estimating the system state from the measurement output that are available is called the estimator design. By utilizing the state estimator, the output feedback problem is converted to the state-feedback problem for a new problem. This new problem employs the estimated state as its own state variable and the solution of the new state-feedback problem leads to the solution of the dynamic output feedback control problem. Basically, the dynamic output feedback is a coupling of control and estimation.

This section aims at designing a full order dynamic $\mathcal{H}_\infty$ fuzzy output feedback controller of the form

$$\begin{aligned} \dot{\hat{x}}(t) &= \textstyle\sum_{i=1}^r \sum_{j=1}^r \hat{\mu}_i \hat{\mu}_j \left[\hat{A}_{ij}\hat{x}(t) + \hat{B}_i y(t)\right] \\ u(t) &= \textstyle\sum_{i=1}^r \hat{\mu}_i \hat{C}_i \hat{x}(t) \end{aligned} \tag{3.12}$$

where $\hat{x}(t) \in \Re^n$ is the controller's state vector, $\hat{A}_{ij}$, $\hat{B}_i$ and $\hat{C}_i$ are parameters of the controller which are to be determined, and $\hat{\mu}_i$ denotes the normalized time-varying fuzzy weighting functions for each rule (i.e., $\hat{\mu}_i \geq 0$ and $\sum_{i=1}^r \hat{\mu}_i = 1$), such that the inequality (3.3) holds.

In this section, we consider the designing of the robust $\mathcal{H}_\infty$ output feedback control into two cases as follows. In Subsection 3.3.1, we consider the case where the premise variable of the fuzzy model μ_i is measurable, while in Subsection 3.3.2, the premise variable which is assumed to be unmeasurable is considered.

3.3.1 Case I–$\nu(t)$ is available for feedback

The premise variable of the fuzzy model $\nu(t)$ is available for feedback which implies that μ_i is available for feedback. Thus, we can select our controller that depends on μ_i as follows:

$$\begin{aligned} \dot{\hat{x}}(t) &= \textstyle\sum_{i=1}^r \sum_{j=1}^r \mu_i \mu_j \left[\hat{A}_{ij}\hat{x}(t) + \hat{B}_i y(t)\right] \\ u(t) &= \textstyle\sum_{i=1}^r \mu_i \hat{C}_i \hat{x}(t). \end{aligned} \tag{3.13}$$

Before presenting our next results, the following lemma is recalled.

Lemma 1. *Consider the system (3.1). Given a prescribed $\mathcal{H}_\infty$ performance γ and a positive constant δ, if there exists a matrix $P = P^T$ satisfying the following linear matrix inequalities:*

$$P > 0 \tag{3.14}$$

$$\begin{pmatrix} A_{cl}^{ij}P + P(A_{cl}^{ij})^T & (*)^T & (*)^T \\ (B_{cl}^{ij})^T & -\gamma^2 I & (*)^T \\ C_{cl}^{ij}P & 0 & -I \end{pmatrix} < 0, \quad i,j = 1,2,\cdots,r \tag{3.15}$$

where

$$A_{cl}^{ij} = \begin{bmatrix} A_i & B_{2_i}\hat{C}_j \\ \hat{B}_i C_{2_j} & \hat{A}_{ij} \end{bmatrix}, \quad B_{cl}^{ij} = \begin{bmatrix} \tilde{B}_{1_i} \\ \hat{B}_i \tilde{D}_{21_j} \end{bmatrix} \quad \textit{and} \quad C_{cl}^{ij} = [\tilde{C}_{1_i} \quad \tilde{D}_{12_i}\hat{C}_j]$$

with

$$\tilde{B}_{1_i} = \left[\delta I\ I\ \delta I\ 0\ B_{1_i}\ 0\right],\ \tilde{C}_{1_i} = \left[\tfrac{\gamma\rho}{\delta}H_{1_i}^T\ 0\ \tfrac{\gamma\rho}{\delta}H_{5_i}^T\ \sqrt{2}\lambda\rho H_{4_i}^T\ \sqrt{2}\lambda C_{1_i}^T\right]^T,$$

$$\tilde{D}_{12_i} = \left[0\ \tfrac{\gamma\rho}{\delta}H_{3_i}^T\ 0\ \sqrt{2}\lambda\rho H_{6_i}^T\ \sqrt{2}\lambda D_{12_i}^T\right]^T,\ \tilde{D}_{21_i} = \left[0\ 0\ 0\ \delta I\ D_{21_i}\ I\right]$$

$$\text{and } \lambda = \left(1 + \rho^2 \sum_{i=1}^{r}\sum_{j=1}^{r}\left[\|H_{2_i}^T H_{2_j}\| + \|H_{7_i}^T H_{7_j}\|\right]\right)^{\frac{1}{2}},$$

then the inequality (3.3) is guaranteed.

Proof: See Appendix. ■

Knowing that the controller's premise variable is the same as the plant's premise variable, the left hand of (3.15) can be re-expressed as follows:

$$A_{cl}^{ij}P + P(A_{cl}^{ij})^T + \gamma^{-2}B_{cl}^{ij}(B_{cl}^{ij})^T + P(C_{cl}^{ij})^T C_{cl}^{ij}P. \tag{3.16}$$

Before providing LMI-based sufficient conditions for the system (3.1) to have an $\mathcal{H}_\infty$ performance, let us partition the matrix P as follows:

$$P = \begin{bmatrix} X & Y^{-1} - X \\ Y^{-1} - X & X - Y^{-1} \end{bmatrix} \tag{3.17}$$

where $X \in \Re^{n\times n}$ and $Y \in \Re^{n\times n}$. Utilizing the partition above, we define the new controller's input and output matrices as

$$\begin{aligned} \mathcal{B}_i &\triangleq \left[Y^{-1} - X\right]\hat{B}_i \\ \mathcal{C}_i &\triangleq \hat{C}_i Y. \end{aligned} \tag{3.18}$$

Using these changes of variable, we have the following theorem.

Theorem 2. *Consider the system (3.1). Given a prescribed $\mathcal{H}_\infty$ performance $\gamma > 0$ and a positive constant δ, if there exist matrices $X = X^T$, $Y = Y^T$, $\mathcal{B}_i$ and $\mathcal{C}_i$, $i = 1, 2, \cdots, r$, satisfying the following linear matrix inequalities:*

$$\begin{bmatrix} X & I \\ I & Y \end{bmatrix} > 0 \tag{3.19}$$

$$X > 0 \tag{3.20}$$

$$Y > 0 \tag{3.21}$$

$$\Psi_{11_{ii}} < 0, \quad i = 1, 2, \cdots, r \tag{3.22}$$

$$\Psi_{22_{ii}} < 0, \quad i = 1, 2, \cdots, r \tag{3.23}$$

$$\Psi_{11_{ij}} + \Psi_{11_{ji}} < 0, \quad i < j \le r \tag{3.24}$$

$$\Psi_{22_{ij}} + \Psi_{22_{ji}} < 0, \quad i < j \le r \tag{3.25}$$

where

$$\Psi_{11_{ij}} = \begin{pmatrix} A_iY + YA_i^T + B_{2_i}\mathcal{C}_j + \mathcal{C}_i^T B_{2_j}^T + \gamma^{-2}\tilde{B}_{1_i}\tilde{B}_{1_j}^T & (*)^T \\ \left[Y\tilde{C}_{1_i}^T + \mathcal{C}_i^T \tilde{D}_{12_j}^T\right]^T & -I \end{pmatrix} \quad (3.26)$$

$$\Psi_{22_{ij}} = \begin{pmatrix} A_i^T X + XA_i + \mathcal{B}_i C_{2_j} + C_{2_i}^T \mathcal{B}_j^T + \tilde{C}_{1_i}^T \tilde{C}_{1_j} & (*)^T \\ \left[X\tilde{B}_{1_i} + \mathcal{B}_i \tilde{D}_{21_j}\right]^T & -\gamma^2 I \end{pmatrix} \quad (3.27)$$

with

$$\tilde{B}_{1_i} = \left[\delta I \; I \; \delta I \; 0 \; B_{1_i} \; 0\right], \quad \tilde{C}_{1_i} = \left[\tfrac{\gamma\rho}{\delta} H_{1_i}^T \; 0 \; \tfrac{\gamma\rho}{\delta} H_{5_i}^T \; \sqrt{2}\lambda\rho H_{4_i}^T \; \sqrt{2}\lambda C_{1_i}^T\right]^T,$$

$$\tilde{D}_{12_i} = \left[0 \; \tfrac{\gamma\rho}{\delta} H_{3_i}^T \; 0 \; \sqrt{2}\lambda\rho H_{6_i}^T \; \sqrt{2}\lambda D_{12_i}^T\right]^T, \quad \tilde{D}_{21_i} = \left[0 \; 0 \; 0 \; \delta I \; D_{21_i} \; I\right]$$

$$\text{and } \lambda = \left(1 + \rho^2 \sum_{i=1}^{r}\sum_{j=1}^{r}\left[\|H_{2_i}^T H_{2_j}\| + \|H_{7_i}^T H_{7_j}\|\right]\right)^{\frac{1}{2}},$$

then the prescribed $\mathcal{H}_\infty$ performance $\gamma > 0$ is guaranteed. Furthermore, a suitable controller is of the form (3.13) with

$$\begin{aligned} \hat{A}_{ij} &= \left[Y^{-1} - X\right]^{-1} \mathcal{M}_{ij} Y^{-1} \\ \hat{B}_i &= \left[Y^{-1} - X\right]^{-1} \mathcal{B}_i \\ \hat{C}_i &= \mathcal{C}_i Y^{-1} \end{aligned} \quad (3.28)$$

where

$$\mathcal{M}_{ij} = -A_i^T - XA_iY - XB_{2_i}\hat{C}_jY - \left[Y^{-1} - X\right]\hat{B}_i C_{2_j} Y$$
$$-\tilde{C}_{1_i}^T\left[\tilde{C}_{1_j}Y + \tilde{D}_{12_j}\hat{C}_jY\right] \quad -\gamma^{-2}\left\{X\tilde{B}_{1_i} + \left[Y^{-1} - X\right]\hat{B}_i\tilde{D}_{21_i}\right\}\tilde{B}_{1_j}^T. \quad (3.29)$$

Proof: Suppose there exist X and Y such that the inequalities (3.19) and (3.20)-(3.21) hold. The inequality (3.19) implies that the matrix P defined in (3.16) is a positive definite matrix. Using the partition (3.17), the controller (3.18) and multiplying (3.16) to the left by $\begin{bmatrix} Y & I \\ Y & 0 \end{bmatrix}$ and to the right by $\begin{bmatrix} Y & Y \\ I & 0 \end{bmatrix}$, we have

$$\begin{bmatrix} \Phi_{11_{ij}} & 0 \\ 0 & \Phi_{22_{ij}} \end{bmatrix} \quad (3.30)$$

where

$$\begin{aligned} \Phi_{11_{ij}} &= A_iY + YA_i^T + B_{2_i}\mathcal{C}_j + \mathcal{C}_i^T B_{2_j}^T + \gamma^{-2}\tilde{B}_{1_i}\tilde{B}_{1_j}^T \\ &\quad + \left[Y\tilde{C}_{1_i}^T + \mathcal{C}_i^T\tilde{D}_{12_j}^T\right]\left[Y\tilde{C}_{1_i}^T + \mathcal{C}_i^T\tilde{D}_{12_j}^T\right]^T \end{aligned} \quad (3.31)$$

$$\begin{aligned} \Phi_{22_{ij}} &= A_i^T X + XA_i + \mathcal{B}_i C_{2_j} + C_{2_i}^T\mathcal{B}_j^T + \tilde{C}_{1_i}^T\tilde{C}_{1_j} \\ &\quad + \gamma^{-2}\left[X\tilde{B}_{1_i} + \mathcal{B}_i\tilde{D}_{21_j}\right]\left[X\tilde{B}_{1_i} + \mathcal{B}_i\tilde{D}_{21_j}\right]^T. \end{aligned} \quad (3.32)$$

Note that $\Phi_{11_{ij}}$ and $\Phi_{22_{ij}}$ are the Schur complements of $\Psi_{11_{ij}}$ and $\Psi_{22_{ij}}$, Using (3.22)-(3.25), we have (3.30) less than zero. Hence, by Theorem 2, we learn that the inequality (3.3) holds. ■

3.3.2 Case II–$\nu(t)$ is unavailable for feedback

The output feedback fuzzy controller is assumed to be the same as the premise variables of the fuzzy system model. This actually means that the premise variables of fuzzy system model are assumed to be measurable. However, in general, it is extremely difficult to derive an accurate fuzzy system model by imposing that all premise variables are measurable. In this subsection, we do not impose that condition, we choose the premise variables of the controller to be different from the premise variables of fuzzy system model of the plant. In here, the premise variables of the controller are selected to be the estimated premise variables of the plant. In the other words, the premise variable of the fuzzy model $\nu(t)$ is unavailable for feedback which implies μ_i is unavailable for feedback. Hence, we cannot select our controller which depends on μ_i. Thus, we select our controller as follows:

$$\begin{aligned} \dot{\hat{x}}(t) &= \textstyle\sum_{i=1}^{r}\sum_{j=1}^{r}\hat{\mu}_i\hat{\mu}_j\Big[\hat{A}_{ij}\hat{x}(t)+\hat{B}_i y(t)\Big] \\ u(t) &= \textstyle\sum_{i=1}^{r}\hat{\mu}_i\hat{C}_i\hat{x}(t). \end{aligned} \tag{3.33}$$

where $\hat{\mu}_i$ depends on the premise variable of the controller which is different from μ_i.

Let us re-express the system (3.1) in terms of $\hat{\mu}_i$, thus the plant's premise variable becomes the same as the controller's premise variable. By doing so, the result given in the previous case can then be applied here. First, let us rewrite (3.1) as follows:

$$\begin{aligned} \dot{x}(t) &= \textstyle\sum_{i=1}^{r}\mu_i\Big[[A_i+\Delta A_i]x(t)+[B_{1_i}+\Delta B_{1_i}]w(t)+[B_{2_i}+\Delta B_{2_i}]u(t)\Big] \\ &\quad +\textstyle\sum_{i=1}^{r}\hat{\mu}_i\Big[[A_i+\Delta A_i]x(t)+[B_{1_i}+\Delta B_{1_i}]w(t)+[B_{2_i}+\Delta B_{2_i}]u(t)\Big] \\ &\quad -\textstyle\sum_{i=1}^{r}\hat{\mu}_i\Big[[A_i+\Delta A_i]x(t)+[B_{1_i}+\Delta B_{1_i}]w(t)+[B_{2_i}+\Delta B_{2_i}]u(t)\Big] \\ z(t) &= \textstyle\sum_{i=1}^{r}\mu_i\Big[[C_{1_i}+\Delta C_{1_i}]x(t)+[D_{12_i}+\Delta D_{12_i}]u(t)\Big] \\ &\quad +\textstyle\sum_{i=1}^{r}\hat{\mu}_i\Big[[C_{1_i}+\Delta C_{1_i}]x(t)+[D_{12_i}+\Delta D_{12_i}]u(t)\Big] \\ &\quad -\textstyle\sum_{i=1}^{r}\hat{\mu}_i\Big[[C_{1_i}+\Delta C_{1_i}]x(t)+[D_{12_i}+\Delta D_{12_i}]u(t)\Big] \\ y(t) &= \textstyle\sum_{i=1}^{r}\mu_i\Big[[C_{2_i}+\Delta C_{2_i}]x(t)+[D_{21_i}+\Delta D_{21_i}]w(t)\Big] \\ &\quad +\textstyle\sum_{i=1}^{r}\hat{\mu}_i\Big[[C_{2_i}+\Delta C_{2_i}]x(t)+[D_{21_i}+\Delta D_{21_i}]w(t)\Big] \\ &\quad -\textstyle\sum_{i=1}^{r}\hat{\mu}_i\Big[[C_{2_i}+\Delta C_{2_i}]x(t)+[D_{21_i}+\Delta D_{21_i}]w(t)\Big]. \end{aligned} \tag{3.34}$$

Rearranging (3.34) together with employing Assumption 3.1, we obtain

$$
\begin{aligned}
\dot{x}(t) = \textstyle\sum_{i=1}^{r} \hat{\mu}_i \Big(& [A_i + F(x(t),t)H_{1_i} + (\mu_1 - \hat{\mu}_1)A_1 + \cdots + (\mu_r - \hat{\mu}_r)A_r \\
& + F(x(t),t)(\mu_1 - \hat{\mu}_1)H_{1_1} + \cdots + F(x(t),t)(\mu_r - \hat{\mu}_r)H_{1_r}]x(t) \\
& + [B_{1_i} + F(x(t),t)H_{2_i} + (\mu_1 - \hat{\mu}_1)B_{1_1} + \cdots + (\mu_r - \hat{\mu}_r)B_{1_r} \\
& + F(x(t),t)(\mu_1 - \hat{\mu}_1)H_{2_1} + \cdots + F(x(t),t)(\mu_r - \hat{\mu}_r)H_{2_r}]w(t) \\
& + [B_{2_i} + F(x(t),t)H_{3_i} + (\mu_1 - \hat{\mu}_1)B_{2_1} + \cdots + (\mu_r - \hat{\mu}_r)B_{2_r} \\
& + F(x(t),t)(\mu_1 - \hat{\mu}_1)H_{3_1} + \cdots + F(x(t),t)(\mu_r - \hat{\mu}_r)H_{3_r}]u(t) \Big) \\
z(t) = \textstyle\sum_{i=1}^{r} \hat{\mu}_i \Big(& [C_{1_i} + F(x(t),t)H_{4_i} + (\mu_1 - \hat{\mu}_1)C_{1_1} + \cdots + (\mu_r - \hat{\mu}_r)C_{1_r} \\
& + F(x(t),t)(\mu_1 - \hat{\mu}_1)H_{4_1} + \cdots + F(x(t),t)(\mu_r - \hat{\mu}_r)H_{4_r}]x(t) \\
& + [D_{12_i} + F(x(t),t)H_{5_i} + (\mu_1 - \hat{\mu}_1)D_{12_1} + \cdots + (\mu_r - \hat{\mu}_r)D_{12_r} \\
& + F(x(t),t)(\mu_1 - \hat{\mu}_1)H_{5_1} + \cdots + F(x(t),t)(\mu_r - \hat{\mu}_r)H_{5_r}]u(t) \Big) \\
y(t) = \textstyle\sum_{i=1}^{r} \hat{\mu}_i \Big(& [C_{2_i} + F(x(t),t)H_{6_i} + (\mu_1 - \hat{\mu}_1)C_{2_1} + \cdots + (\mu_r - \hat{\mu}_r)C_{2_r} \\
& + F(x(t),t)(\mu_1 - \hat{\mu}_1)H_{6_1} + \cdots + F(x(t),t)(\mu_r - \hat{\mu}_r)H_{6_r}]x(t) \\
& + [D_{21_i} + F(x(t),t)H_{7_i} + (\mu_1 - \hat{\mu}_1)D_{21_1} + \cdots + (\mu_r - \hat{\mu}_r)D_{21_r} \\
& + F(x(t),t)(\mu_1 - \hat{\mu}_1)H_{7_1} + \cdots + F(x(t),t)(\mu_r - \hat{\mu}_r)H_{7_r}]w(t) \Big)
\end{aligned}
\tag{3.35}
$$

Then, from (3.35), we get

$$
\begin{aligned}
\dot{x}(t) &= \textstyle\sum_{i=1}^{r} \hat{\mu}_i \Big[[A_i + \Delta\bar{A}_i]x(t) + [B_{1_i} + \Delta\bar{B}_{1_i}]w(t) \\
&\qquad + [B_{2_i} + \Delta\bar{B}_{2_i}]u(t) \Big], \quad x(0) = 0 \\
z(t) &= \textstyle\sum_{i=1}^{r} \hat{\mu}_i \Big[[C_{1_i} + \Delta\bar{C}_{1_i}]x(t) + [D_{12_i} + \Delta\bar{D}_{12_i}]u(t) \Big] \\
y(t) &= \textstyle\sum_{i=1}^{r} \hat{\mu}_i \Big[[C_{2_i} + \Delta\bar{C}_{2_i}]x(t) + [D_{21_i} + \Delta\bar{D}_{21_i}]w(t) \Big]
\end{aligned}
\tag{3.36}
$$

where

$$
\Delta\bar{A}_i = \bar{F}(x,\hat{x},t)\bar{H}_{1_i}, \quad \Delta\bar{B}_{1_i} = \bar{F}(x,\hat{x},t)\bar{H}_{2_i}, \quad \Delta\bar{B}_{2_i} = \bar{F}(x,\hat{x},t)\bar{H}_{3_i},
$$
$$
\Delta\bar{C}_{1_i} = \bar{F}(x,\hat{x},t)\bar{H}_{4_i}, \quad \Delta\bar{C}_{2_i} = \bar{F}(x,\hat{x},t)\bar{H}_{5_i}, \quad \Delta\bar{D}_{12_i} = \bar{F}(x,\hat{x},t)\bar{H}_{6_i}
$$
$$
\text{and } \Delta\bar{D}_{21_i} = \bar{F}(x,\hat{x},t)\bar{H}_{7_i}
$$

with

$$
\bar{H}_{1_i} = \left[H_{1_i}^T \; A_1^T \; \cdots A_r^T \; H_{1_1}^T \cdots H_{1_r}^T\right]^T, \; \bar{H}_{2_i} = \left[H_{2_i}^T \; B_{1_1}^T \cdots B_{1_r}^T \; H_{2_1}^T \cdots H_{2_r}^T\right]^T,
$$
$$
\bar{H}_{3_i} = \left[H_{3_i}^T \; B_{2_1}^T \cdots B_{2_r}^T \; H_{3_1}^T \cdots H_{3_r}^T\right]^T, \; \bar{H}_{4_i} = \left[H_{4_i}^T \; C_{1_1}^T \cdots C_{1_r}^T \; H_{4_1}^T \cdots H_{4_r}^T\right]^T,
$$
$$
\bar{H}_{5_i} = \left[H_{5_i}^T \; C_{2_1}^T \cdots C_{2_r}^T \; H_{5_1}^T \cdots H_{5_r}^T\right]^T, \bar{H}_{6_i} = \left[H_{6_i}^T \; D_{12_1}^T \cdots D_{12_r}^T \; H_{6_1}^T \cdots H_{6_r}^T\right]^T
$$
$$
\bar{H}_{7_i} = \left[H_{7_i}^T \; D_{21_1}^T \cdots \; D_{21_r}^T \; H_{7_1}^T \cdots H_{7_r}^T\right]^T \text{ and}
$$

$\bar{F}(x(t),\hat{x}(t),t) = \Big[F(x(t),t) \;\; (\mu_1 - \hat{\mu}_1) \;\; \cdots \;\; (\mu_r - \hat{\mu}_r) \;\; F(x(t),t)(\mu_1 - \hat{\mu}_1) \;\; \cdots \;\; F(x(t),t)(\mu_r - \hat{\mu}_r)\Big]$. Note that $\|\bar{F}(x(t),\hat{x}(t),t)\| \leq \bar{\rho}$ where $\bar{\rho} = \{3\rho^2 + 2\}^{\frac{1}{2}}$.

$\bar{\rho}$ is derived by utilizing the concept of vector norm in basic system control theory and the fact that $\mu_i \geq 0$, $\hat{\mu}_i \geq 0$, $\sum_{i=1}^{r} \mu_i = 1$ and $\sum_{i=1}^{r} \hat{\mu}_i = 1$.

Note that the above technique is basically employed in order to obtain the plant's premise variable to be the same as the controller's premise variable; e.g. [27, 28]. Now, the premise variable of the system is the same as the premise variable of the controller, thus we can apply the result given in Case I.

Theorem 3. *Consider the system (3.1). Given a prescribed $\mathcal{H}_\infty$ performance $\gamma > 0$ and a positive constant δ, if there exist matrices X, Y, $\mathcal{B}_i$ and $\mathcal{C}_i$, $i = 1, 2, \cdots, r$, satisfying the following linear matrix inequalities:*

$$\begin{bmatrix} X & I \\ I & Y \end{bmatrix} > 0 \tag{3.37}$$

$$X > 0 \tag{3.38}$$

$$Y > 0 \tag{3.39}$$

$$\Psi 11_{ii} < 0, \quad i = 1, 2, \cdots, r \tag{3.40}$$

$$\Psi 22_{ii} < 0, \quad i = 1, 2, \cdots, r \tag{3.41}$$

$$\Psi 11_{ij} + \Psi 11_{ji} < 0, \quad i < j \leq r \tag{3.42}$$

$$\Psi 22_{ij} + \Psi 22_{ji} < 0, \quad i < j \leq r \tag{3.43}$$

where

$$\Psi 11_{ij} = \begin{pmatrix} A_i Y + Y A_i^T + B_{2_i} \mathcal{C}_j + \mathcal{C}_j^T B_{2_i}^T + \gamma^{-2} \tilde{\bar{B}}_{1_i} \tilde{\bar{B}}_{1_j}^T & (*)^T \\ \left[Y \tilde{\bar{C}}_{1_i}^T + \mathcal{C}_j^T \tilde{\bar{D}}_{12_j}^T \right]^T & -I \end{pmatrix} \tag{3.44}$$

$$\Psi 22_{ij} = \begin{pmatrix} A_i^T X + X A_i + \mathcal{B}_i C_{2_j} + C_{2_i}^T \mathcal{B}_j^T + \tilde{\bar{C}}_{1_i}^T \tilde{\bar{C}}_{1_j} & (*)^T \\ \left[X \tilde{\bar{B}}_{1_i} + \mathcal{B}_i \tilde{\bar{D}}_{21_j} \right]^T & -\gamma^2 I \end{pmatrix} \tag{3.45}$$

with

$$\tilde{\bar{B}}_{1_i} = \left[\delta I \; I \; \delta I \; 0 \; B_{1_i} \; 0 \right], \quad \tilde{\bar{C}}_{1_i} = \left[\tfrac{\gamma\bar{\rho}}{\delta} H_{1_i}^T \; 0 \; \tfrac{\gamma\bar{\rho}}{\delta} H_{5_i}^T \; \sqrt{2}\bar{\lambda}\bar{\rho} H_{4_i}^T \; \sqrt{2}\bar{\lambda} C_{1_i}^T \right]^T,$$

$$\tilde{\bar{D}}_{12_i} = \left[0 \; \tfrac{\gamma\bar{\rho}}{\delta} H_{3_i}^T \; 0 \; \sqrt{2}\bar{\lambda}\bar{\rho} H_{6_i}^T \; \sqrt{2}\bar{\lambda} D_{12_i}^T \right]^T, \quad \tilde{\bar{D}}_{21_i} = \left[0 \; 0 \; 0 \; \delta I \; D_{21_i} \; I \right]$$

$$\text{and } \bar{\lambda} = \left(1 + \bar{\rho}^2 \sum_{i=1}^{r} \sum_{j=1}^{r} \left[\| \bar{H}_{2_i}^T \bar{H}_{2_j} \| + \| \bar{H}_{7_i}^T \bar{H}_{7_j} \| \right] \right)^{\frac{1}{2}},$$

then the prescribed $\mathcal{H}_\infty$ performance $\gamma > 0$ is guaranteed. Furthermore, a suitable controller is of the form (3.33) with

$$\begin{aligned} \hat{A}_{ij} &= \left[Y^{-1} - X \right]^{-1} \mathcal{M}_{ij} Y^{-1} \\ \hat{B}_i &= \left[Y^{-1} - X \right]^{-1} \mathcal{B}_i \\ \hat{C}_i &= \mathcal{C}_i Y^{-1} \end{aligned} \tag{3.46}$$

where

$$\begin{aligned}\mathcal{M}_{ij} = &-A_i^T - XA_iY - XB_{2_i}\hat{C}_jY - \left[Y^{-1} - X\right]\hat{B}_iC_{2_j}Y \\ &-\tilde{\tilde{C}}_{1_i}^T\left[\tilde{\tilde{C}}_{1_j}Y + \tilde{\tilde{D}}_{12_j}\hat{C}_jY\right] - \gamma^{-2}\left\{X\tilde{\tilde{B}}_{1_i} + \left[Y^{-1} - X\right]\hat{B}_i\tilde{\tilde{D}}_{21_i}\right\}\tilde{\tilde{B}}_{1_j}^T.\end{aligned}$$

Proof: Since (3.36) is of the form of (3.1), it can be shown by employing the proof for Theorem 2. ■

3.4 Example

Consider the following problem of the chaotic Lorenz system which is described by the following equations; see [59]

$$\begin{aligned}\dot{x}_1(t) &= -\sigma x_1(t) + \sigma x_2(t) + u(t) + 0.1w_1(t) \\ \dot{x}_2(t) &= rx_1(t) - x_2(t) - x_1(t)x_3(t) + 0.1w_2(t) \\ \dot{x}_3(t) &= x_1(t)x_2(t) - bx_3(t) + 0.1w_3(t) \\ z(t) &= \left[x_1^T(t)\ x_2^T(t)\ x_3^T(t)\right]^T \\ y(t) &= Jx(t) + 0.1w_1(t)\end{aligned} \tag{3.47}$$

where $x_1(t)$, $x_2(t)$, $x_3(t)$ denote the state vectors, $u(t)$ is the control input, $w_1(t)$, $w_2(t)$, $w_3(t)$ are the disturbance noise inputs, $y(t)$ is the measurement output, $z(t)$ is the controlled output, J is the sensor matrix and the bounded uncertain parameters σ, r and b are given by $10 \pm 30\%$, $28 \pm 30\%$ and $8/3 \pm 30\%$, respectively. Note that the variables $x_1(t)$, $x_2(t)$ and $x_3(t)$ are treated as the deviation variables (variables deviate from the desired trajectories).

Since the nonlinear terms in (3.47) can be viewed as a function of $x_1(t)$, we can re-expressed (3.47) as

$$\begin{aligned}\dot{x}_1(t) &= -\sigma x_1(t) + \sigma x_2(t) + u(t) + 0.1w_1(t) \\ \dot{x}_2(t) &= rx_1(t) - x_2(t) - (x_1(t)) \cdot x_3(t) + 0.1w_2(t) \\ \dot{x}_3(t) &= (x_1(t)) \cdot x_2(t) - bx_3(t) + 0.1w_3(t) \\ z(t) &= \left[x_1^T(t)\ x_2^T(t)\ x_3^T(t)\right]^T \\ y(t) &= Jx(t) + 0.1w_1(t).\end{aligned} \tag{3.48}$$

The control objective is to control the state variable $x_1(t)$ for the range $x_1(t) \in [N_1\ N_2]$. For the sake of simplicity, we will use as few rules as possible. Note that Figure 3.1 shows the plot of the membership functions represented by

$$M_1(x_1(t)) = \frac{-x_1(t) + N_2}{N_2 - N_1} \quad \text{and} \quad M_2(x_1(t)) = \frac{x_1(t) - N_1}{N_2 - N_1}.$$

Knowing that $x_1(t) \in [N_1\ N_2]$, the nonlinear system (3.48) can be approximated by the following two rules TS model:

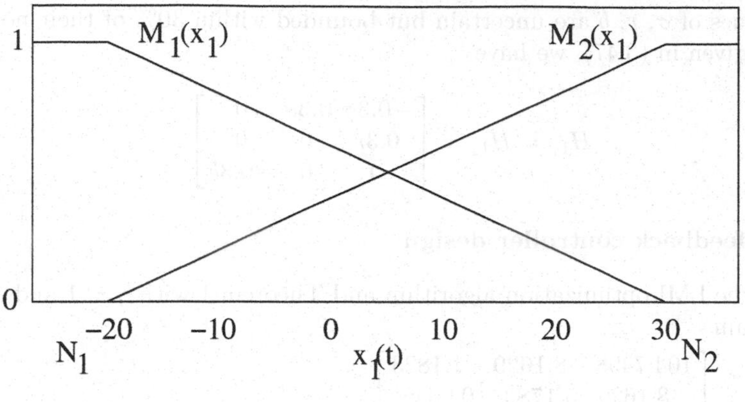

Fig. 3.1. Membership functions for the two fuzzy set.

Plant Rule 1: IF $x_1(t)$ is $M_1(x_1(t))$ THEN

$$\begin{aligned}\dot{x}(t) &= [A_1 + \Delta A_1]x(t) + B_{1_1}w(t) + B_{2_1}u(t), \quad x(0) = 0,\\ z(t) &= C_{1_1}x(t),\\ y(t) &= C_{2_1}x(t) + D_{21_1}w(t).\end{aligned}$$

Plant Rule 2: IF $x_1(t)$ is $M_2(x_1(t))$ THEN

$$\begin{aligned}\dot{x}(t) &= [A_2 + \Delta A_2]x(t) + B_{1_2}w(t) + B_{2_2}u(t), \quad x(0) = 0,\\ z(t) &= C_{1_2}x(t),\\ y(t) &= C_{2_2}x(t) + D_{21_2}w(t)\end{aligned}$$

where

$$A_1 = \begin{bmatrix} -10 & 10 & 0 \\ 28 & -1 & -N_1 \\ 0 & N_1 & -8/3 \end{bmatrix}, \quad A_2 = \begin{bmatrix} -10 & 10 & 0 \\ 28 & -1 & -N_2 \\ 0 & N_2 & -8/3 \end{bmatrix},$$

$$B_{1_1} = B_{1_2} = \begin{bmatrix} 0.1 & 0 & 0 \\ 0 & 0.1 & 0 \\ 0 & 0 & 0.1 \end{bmatrix}, \quad B_{2_1} = B_{2_2} \begin{bmatrix} 1 \\ 0 \\ 0 \end{bmatrix},$$

$$C_{1_1} = C_{1_2} \begin{bmatrix} 1 & 0 & 0 \\ 0 & 1 & 0 \\ 0 & 0 & 1 \end{bmatrix}, \quad C_{2_1} = C_{2_2} = J, \quad D_{21_1} = D_{21_2} = \begin{bmatrix} 0.1 & 0 & 0 \end{bmatrix},$$

$$\Delta A_1 = F(x(t), t)H_{1_1}, \quad \Delta A_2 = F(x(t), t)H_{1_2},$$

$$x(t) = [x_1^T(t)\ x_2^T(t)\ x_3^T(t)]^T \text{ and } w(t) = [w_1^T(t)\ w_2^T(t)\ w_3^T(t)]^T.$$

Let us choose the value of $[N_1 \quad N_2]$ in the membership function as $[-20 \quad 30]$. Now, by assuming that in (3.2), $\|F(x(t), t)\| \leq \rho = 1$ and since the values of σ, r, b are uncertain but bounded within 30% of their nominal values given in (3.47), we have

$$H_{1_1} = H_{1_2} = \begin{bmatrix} -0.3\sigma & 0.3\sigma & 0 \\ 0.3r & 0 & 0 \\ 0 & 0 & -0.3b \end{bmatrix}.$$

State-feedback controller design

Using the LMI optimization algorithm and Theorem 1 with $\gamma = 1$ and $\delta = 1$, we obtain

$$P = \begin{bmatrix} 104.7498 & -8.1629 & -1.1823 \\ -8.1629 & 5.1783 & 0.9345 \\ -1.1823 & 0.9345 & 6.7383 \end{bmatrix},$$

$$K_1 = \begin{bmatrix} -38.8875 & -816.1115 & -3.9273 \end{bmatrix},$$

$$K_2 = \begin{bmatrix} -37.4290 & -815.5695 & 4.1287 \end{bmatrix}.$$

The resulting fuzzy controller is

$$u(t) = \sum_{j=1}^{2} \mu_j K_j x(t)$$

where

$$\mu_1 = M_1(x_1(t)) \text{ and } \mu_2 = M_2(x_1(t)).$$

Output feedback controller design

Case I: $\nu(t)$ are available for feedback

In this case, $x_1(t) = \nu(t)$ is assumed to be available for feedback; for instance, $J = [1\ 0\ 0]$. This implies that μ_i is available for feedback. Using the LMI optimization algorithm and Theorem 2 with $\gamma = 1$ and $\delta = 1$, we obtain the following results:

$$X = \begin{bmatrix} 40.961 & -0.3001 & 0.0003 \\ -0.3001 & 0.0326 & -0.0020 \\ 0.0003 & -0.0020 & 0.0529 \end{bmatrix}, Y = \begin{bmatrix} 64.041 & -6.6279 & -0.0180 \\ -6.6279 & 0.7784 & 0.0345 \\ -0.0180 & 0.0345 & 0.8385 \end{bmatrix},$$

$$\hat{A}_{11} = \begin{bmatrix} -52.645 & 913.03 & 11.1683 \\ 0.4211 & -93.811 & -1.1292 \\ 2.3239 & -0.4233 & 0.0865 \end{bmatrix}, \hat{A}_{12} = \begin{bmatrix} -52.974 & 909.63 & 0.8313 \\ 0.5070 & -93.0535 & -0.2157 \\ 2.3414 & -0.2540 & 0.1024 \end{bmatrix},$$

$$\hat{A}_{21} = \begin{bmatrix} -54.839 & 912.45 & -6.7553 \\ 1.4467 & -93.619 & 0.6829 \\ -3.5367 & -0.1599 & 0.2080 \end{bmatrix}, \hat{A}_{22} = \begin{bmatrix} -54.767 & 913.461 & -17.1638 \\ 1.3897 & -94.074 & 1.5985 \\ -3.5229 & -0.0374 & 0.1865 \end{bmatrix},$$

$$\hat{B}_1 = \begin{bmatrix} -110.4306 \\ 4.8589 \\ 2.9909 \end{bmatrix}, \hat{B}_2 = \begin{bmatrix} 113.2188 \\ 6.1387 \\ -4.5464 \end{bmatrix},$$

$$\hat{C}_1 = \begin{bmatrix} -36.1488 & -710.9845 & -3.2817 \end{bmatrix}, \hat{C}_2 = \begin{bmatrix} -35.9847 & -709.7215 & 5.1803 \end{bmatrix}.$$

The resulting fuzzy controller is

$$\dot{\hat{x}}(t) = \sum_{i=1}^{2}\sum_{j=1}^{2} \mu_i\mu_j\hat{A}_{ij}\hat{x}(t) + \sum_{i=1}^{2}\mu_i\hat{B}_i y(t)$$

$$u(t) = \sum_{i=1}^{2}\mu_i\hat{C}_i\hat{x}(t)$$

where

$$\mu_1 = M_1(x_1(t)) \text{ and } \mu_2 = M_2(x_1(t)).$$

Case II: $\nu(t)$ are unavailable for feedback

In this case, $x_1(t) = \nu(t)$ is assumed to be unavailable for feedback; for instance, $J = [0\ 0\ 1]$. This implies that μ_i is unavailable for feedback. Using the LMI optimization algorithm and Theorem 3 with $\gamma = 1$ and $\delta = 1$, we obtain the following results:

$$X = \begin{bmatrix} 15.386 & -0.0454 & 0.0001 \\ -0.0454 & 0.0086 & -0.0005 \\ 0.0001 & -0.0005 & 0.0121 \end{bmatrix}, Y = \begin{bmatrix} 195.08 & -19.857 & -0.0836 \\ -19.857 & 2.3203 & 0.1018 \\ -0.0836 & 0.1018 & 2.5038 \end{bmatrix},$$

$$\hat{A}_{11} = \begin{bmatrix} -72.511 & 1594.5 & 6.3456 \\ 5.0232 & -162.66 & -0.6001 \\ 1.2000 & -0.7556 & 0.1000 \end{bmatrix}, \hat{A}_{12} = \begin{bmatrix} -72.923 & 1603.7 & -9.7233 \\ 5.1345 & -162.85 & 0.9974 \\ 1.2000 & -0.5689 & 0.1000 \end{bmatrix},$$

$$\hat{A}_{21} = \begin{bmatrix} -74.545 & 1595.2 & -5.6743 \\ 5.5411 & -162.17 & 0.5609 \\ -1.7009 & -0.9421 & 0.2000 \end{bmatrix}, \hat{A}_{22} = \begin{bmatrix} -74.529 & 1595.2 & -5.6744 \\ 5.5411 & -162.13 & 0.5966 \\ -1.7008 & -0.9432 & 0.2000 \end{bmatrix},$$

$$\hat{B}_1 = \begin{bmatrix} -166.7783 \\ 7.4682 \\ 4.5048 \end{bmatrix}, \hat{B}_2 = \begin{bmatrix} -173.8473 \\ 9.1193 \\ -6.8346 \end{bmatrix},$$

$$\hat{C}_1 = \begin{bmatrix} 14.193 & -410.52 & -0.3593 \end{bmatrix}, \hat{C}_2 = \begin{bmatrix} 14.236 & -412.97 & 3.8984 \end{bmatrix}.$$

The resulting fuzzy controller is

$$\dot{\hat{x}}(t) = \sum_{i=1}^{2}\sum_{j=1}^{2} \hat{\mu}_i\hat{\mu}_j\hat{A}_{ij}\hat{x}(t) + \sum_{i=1}^{2}\hat{\mu}_i\hat{B}_i y(t)$$

$$u(t) = \sum_{i=1}^{2}\hat{\mu}_i\hat{C}_i\hat{x}(t)$$

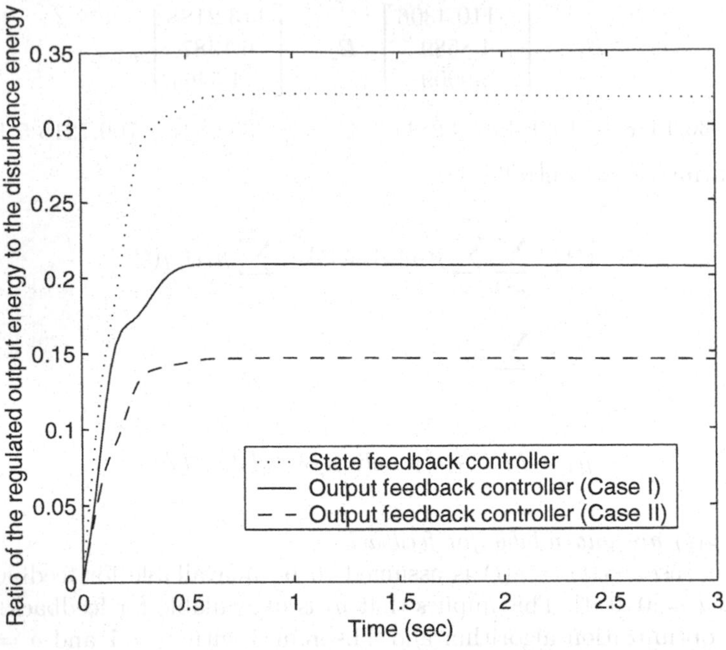

Fig. 3.2. The ratio of the regulated output energy to the disturbance noise energy: $\left(\frac{\int_0^{T_f} z^T(t)z(t)dt}{\int_0^{T_f} w^T(t)w(t)dt}\right)$.

where

$$\hat{\mu}_1 = M_1(\hat{x}_1(t)) \text{ and } \hat{\mu}_2 = M_2(\hat{x}_1(t)).$$

Remark 1. Both robust fuzzy state and output controllers guarantee that the $\mathcal{L}_2$-gain, γ, is less than the prescribed value. The ratio of the regulated output energy to the disturbance input noise energy which is obtained by using the $\mathcal{H}_\infty$ fuzzy controllers is depicted in Figure 3.2. The disturbance input signals, $w_1(t)$, $w_2(t)$ and $w_3(t)$, which were used during the simulation is given in Figure 3.3. After 3 *seconds, the ratio of the regulated output energy to the disturbance input noise energy tends to a constant value which is about* 0.32 *for the state-feedback controller, and* 0.21 *for the output feedback controller in Case I and* 0.14 *in Case II. Thus, for the state-feedback controller where $\gamma = \sqrt{0.32} = 0.566$, for output feedback controller in Case I where $\gamma = \sqrt{0.21} = 0.458$ and in Case II where $\gamma = \sqrt{0.14} = 0.374$, all are less than the prescribed value* 1. □

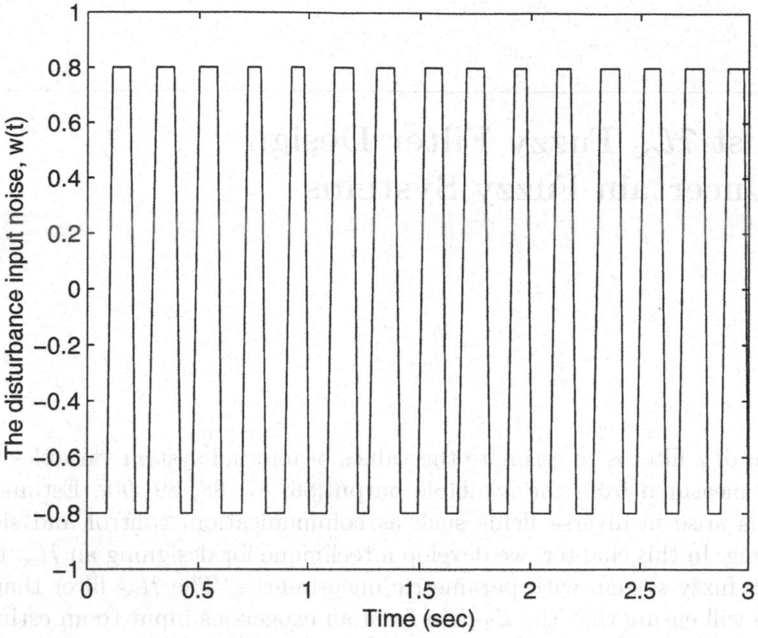

Fig. 3.3. The disturbance input signals, $w_1(t)$, $w_2(t)$ and $w_3(t)$.

3.5 Conclusion

This chapter has investigated the problem of designing a robust fuzzy controller for a TS fuzzy system with parametric uncertainties that guarantees the $\mathcal{L}_2$-gain from an exogenous input to a regulated output being less than or equal to the prescribed value. An LMI-based approach has been employed to derive sufficient conditions for the existence of a robust $\mathcal{H}_\infty$ fuzzy controller in terms of a family of LMIs. Finally, a numerical simulation example has been presented to illustrate the effectiveness of the designs.

4

Robust $\mathcal{H}_\infty$ Fuzzy Filter Design for Uncertain Fuzzy Systems

The aim of a filter is to estimate the values of internal system variables that are not measured from the available output[86, 87, 88, 89, 90]. Estimation problems arise in diverse fields such as communication, control and signal processing. In this chapter, we develop a technique for designing an $\mathcal{H}_\infty$ filter for a TS fuzzy system with parametric uncertainties. The $\mathcal{H}_\infty$ filter that we propose will ensure that the $\mathcal{L}_2$-gain from an exogenous input to an estimate error output is less than or equal to a prescribed value.

4.1 Robust $\mathcal{H}_\infty$ Fuzzy Filter Design

This chapter deals with the problem of designing an $\mathcal{H}_\infty$ filter. Without loss of generality, we assume that $u(t) = 0$. Let us recall the system (3.1) with $u(t) = 0$ as follows:

$$\begin{aligned} \dot{x}(t) &= \sum_{i=1}^{r} \mu_i \Big[[A_i + \Delta A_i]x(t) + [B_{1_i} + \Delta B_{1_i}]w(t)\Big], \ \ x(0) = 0 \\ z(t) &= \sum_{i=1}^{r} \mu_i \Big[[C_{1_i} + \Delta C_{1_i}]x(t)\Big] \\ y(t) &= \sum_{i=1}^{r} \mu_i \Big[[C_{2_i} + \Delta C_{2_i}]x(t) + [D_{21_i} + \Delta D_{21_i}]w(t)\Big]. \end{aligned} \tag{4.1}$$

We are now aiming to design a full order dynamic $\mathcal{H}_\infty$ fuzzy filter of the form

$$\begin{aligned} \dot{\hat{x}}(t) &= \sum_{i=1}^{r} \sum_{j=1}^{r} \hat{\mu}_i \hat{\mu}_j \Big[\hat{A}_{ij}\hat{x}(t) + \hat{B}_i y(t)\Big] \\ \hat{z}(t) &= \sum_{i=1}^{r} \hat{\mu}_i \hat{C}_i \hat{x}(t) \end{aligned} \tag{4.2}$$

where $\hat{x}(t) \in \Re^n$ is the filter's state vector, $\hat{z} \in \Re^s$ is the estimate of $z(t)$, $\hat{A}_{ij}$, $\hat{B}_i$ and $\hat{C}_i$ are parameters of the filter which are to be determined, and $\hat{\mu}_i$ denotes the normalized time-varying fuzzy weighting functions for each rule (i.e., $\hat{\mu}_i \geq 0$ and $\sum_{i=1}^{r} \hat{\mu}_i = 1$), such that the following inequality holds

$$\int_0^{T_f} \Big(z(t) - \hat{z}(t)\Big)^T \Big(z(t) - \hat{z}(t)\Big) dt \le \gamma^2 \left[\int_0^{T_f} w^T(t)w(t)dt\right] \quad (4.3)$$

with $x(0) = 0$, where $(z(t) - \hat{z}(t))$ is the estimated error output, for all $T_f \ge 0$ and $w(t) \in \mathcal{L}_2[0, T_f]$.

Figure 4.1 shows the block diagram of a robust fuzzy filtering problem associated with an uncertain fuzzy system. The major implication of this approach is that the structure of the filter has to take into a account the effect of uncertainty. The problem addressed is the design of a filter such that the induced operator norm of the mapping from the noise $w(t)$ to the filter error $e(t) = z(t) - \hat{z}(t)$ is kept within a prescribed bound for all admissible parameter uncertainties. Clearly, in real control problems, all of the premise variables are not necessarily measurable, thus two cases will be considered in this section. Subsection 4.1.1 considers the case where the premise variable of the fuzzy model μ_i is measurable, while in Subsection 4.1.2, the premise variable is assumed to be unmeasurable.

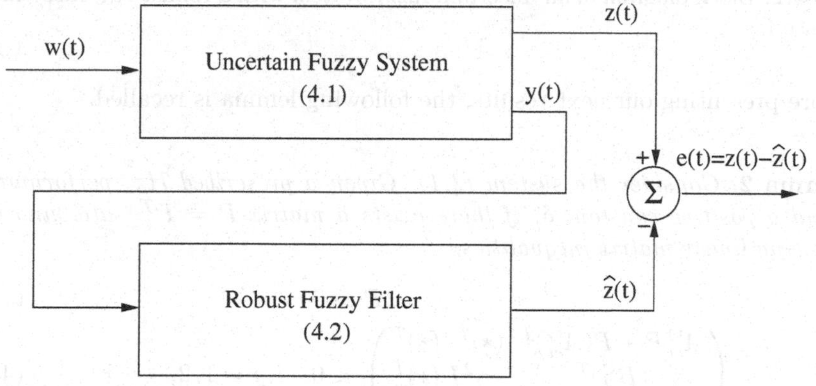

Fig. 4.1. Block diagram of an uncertain fuzzy system with a robust $\mathcal{H}_\infty$ fuzzy filter.

4.1.1 Case I–$\nu(t)$ is available for feedback

The premise variable of the fuzzy model $\nu(t)$ is available for feedback which implies that μ_i is available for feedback. Thus, we can select our filter that depends on μ_i as follows [91]:

$$\begin{aligned} \dot{\hat{x}}(t) &= \textstyle\sum_{i=1}^r \sum_{j=1}^r \mu_i\mu_j \Big[\hat{A}_{ij}\hat{x}(t) + \hat{B}_i y(t)\Big] \\ \hat{z}(t) &= \textstyle\sum_{i=1}^r \mu_i \hat{C}_i \hat{x}(t). \end{aligned} \quad (4.4)$$

Figure 4.2 shows the block diagram of the robust $\mathcal{H}_\infty$ filtering problem associated with uncertain fuzzy system in case that μ_i is available for feedback.

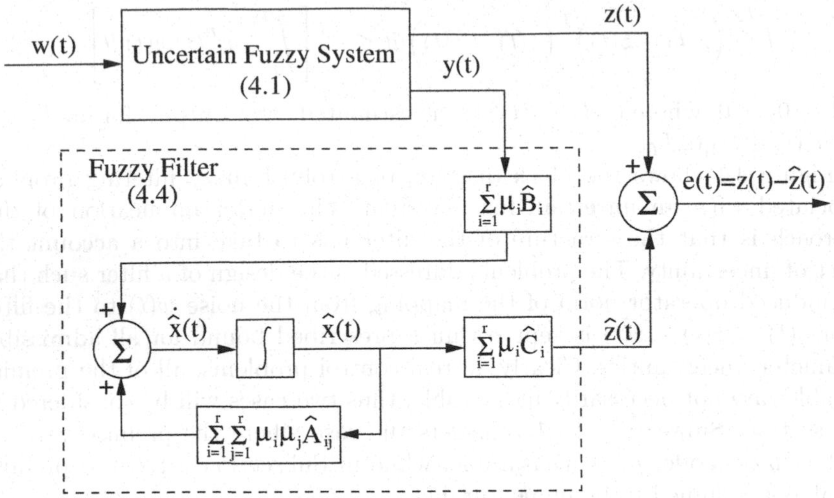

Fig. 4.2. Block diagram of an uncertain fuzzy system with a robust $\mathcal{H}_\infty$ fuzzy filter in Case I.

Before presenting our next results, the following lemma is recalled.

Lemma 2. *Consider the system (4.1). Given a prescribed $\mathcal{H}_\infty$ performance γ and a positive constant δ, if there exists a matrix $P = P^T$ satisfying the following linear matrix inequalities:*

$$P > 0 \tag{4.5}$$

$$\begin{pmatrix} A_{cl}^{ij}P + P(A_{cl}^{ij})^T & (*)^T & (*)^T \\ (B_{cl}^{ij})^T & -\gamma^2 I & (*)^T \\ C_{cl}^{ij}P & 0 & -I \end{pmatrix} < 0, \quad i,j = 1,2,\cdots,r \tag{4.6}$$

where

$$A_{cl}^{ij} = \begin{bmatrix} A_i & 0 \\ \hat{B}_i C_{2_j} & \hat{A}_{ij} \end{bmatrix}, \quad B_{cl}^{ij} = \begin{bmatrix} \tilde{B}_{1_i} \\ \hat{B}_i \tilde{D}_{21_j} \end{bmatrix} \quad \text{and} \quad C_{cl}^{ij} = [\tilde{C}_{1_i} \quad \tilde{D}_{12_i}\hat{C}_j]$$

with

$$\tilde{B}_{1_i} = \begin{bmatrix} \delta I \; I \; 0 \; B_{1_i} \; 0 \end{bmatrix}, \quad \tilde{C}_{1_i} = \begin{bmatrix} \frac{\gamma\rho}{\delta}H_{1_i}^T \; \frac{\gamma\rho}{\delta}H_{5_i}^T \; \sqrt{2}\lambda\rho H_{4_i}^T \; \sqrt{2}\lambda C_{1_i}^T \end{bmatrix}^T,$$

$$\tilde{D}_{12} = \begin{bmatrix} 0 \; 0 \; 0 \; -\sqrt{2}\lambda I \end{bmatrix}^T, \quad \tilde{D}_{21_i} = \begin{bmatrix} 0 \; 0 \; \delta I \; D_{21_i} \; I \end{bmatrix}$$

$$\text{and } \lambda = \left(1 + \rho^2 \sum_{i=1}^{r}\sum_{j=1}^{r}\left[\|H_{2_i}^T H_{2_j}\| + \|H_{7_i}^T H_{7_j}\|\right]\right)^{\frac{1}{2}},$$

then the inequality (4.3) is guaranteed.

Proof: The proof can be carried out by the same technique used in Lemma 3.1. ∎

Knowing that the filter's premise variable is the same as the plant's premise variable, the left hand of (4.6) can be re-expressed as follows:

$$A_{cl}^{ij}P + P(A_{cl}^{ij})^T + \gamma^{-2}B_{cl}^{ij}(B_{cl}^{ij})^T + P(C_{cl}^{ij})^T C_{cl}^{ij} P. \tag{4.7}$$

Before providing LMI-based sufficient conditions for the system (4.1) to have an $\mathcal{H}_\infty$ performance, let us partition the matrix P as follows:

$$P = \begin{bmatrix} X & Y^{-1} - X \\ Y^{-1} - X & X - Y^{-1} \end{bmatrix} \tag{4.8}$$

where $X \in \Re^{n\times n}$ and $Y \in \Re^{n\times n}$. Utilizing the partition above, we define the new filter's input and output matrices as

$$\begin{aligned} \mathcal{B}_i &\triangleq \left[Y^{-1} - X\right]\hat{B}_i \\ \mathcal{C}_i &\triangleq \hat{C}_i Y. \end{aligned} \tag{4.9}$$

Using these changes of variable, we have the following theorem.

Theorem 4. *Consider the system (4.1). Given a prescribed $\mathcal{H}_\infty$ performance $\gamma > 0$ and a positive constant δ, if there exist matrices $X = X^T$, $Y = Y^T$, $\mathcal{B}_i$ and $\mathcal{C}_i$, $i = 1, 2, \cdots, r$, satisfying the following linear matrix inequalities:*

$$\begin{bmatrix} X & I \\ I & Y \end{bmatrix} > 0 \tag{4.10}$$

$$X > 0 \tag{4.11}$$

$$Y > 0 \tag{4.12}$$

$$\Psi_{11_{ii}} < 0, \quad i = 1, 2, \cdots, r \tag{4.13}$$

$$\Psi_{22_{ii}} < 0, \quad i = 1, 2, \cdots, r \tag{4.14}$$

$$\Psi_{11_{ij}} + \Psi_{11_{ji}} < 0, \quad i < j \leq r \tag{4.15}$$

$$\Psi_{22_{ij}} + \Psi_{22_{ji}} < 0, \quad i < j \leq r \tag{4.16}$$

where

$$\Psi_{11_{ij}} = \begin{pmatrix} A_iY + YA_i^T + \gamma^{-2}\tilde{B}_{1_i}\tilde{B}_{1_j}^T & (*)^T \\ \left[Y\tilde{C}_{1_i}^T + \mathcal{C}_i^T\tilde{D}_{12}^T\right]^T & -I \end{pmatrix} \tag{4.17}$$

$$\Psi_{22_{ij}} = \begin{pmatrix} \begin{pmatrix} A_i^TX + XA_i + \mathcal{B}_iC_{2_j} \\ +C_{2_i}^T\mathcal{B}_j^T + \tilde{C}_{1_i}^T\tilde{C}_{1_j} \end{pmatrix} & (*)^T \\ \left[X\tilde{B}_{1_i} + \mathcal{B}_i\tilde{D}_{21_j}\right]^T & -\gamma^2 I \end{pmatrix} \tag{4.18}$$

with

$$\tilde{B}_{1_i} = \begin{bmatrix} \delta I \; I \; 0 \; B_{1_i} \; 0 \end{bmatrix}, \quad \tilde{C}_{1_i} = \begin{bmatrix} \frac{\gamma\rho}{\delta} H_{1_i}^T \; \frac{\gamma\rho}{\delta} H_{5_i}^T \; \sqrt{2}\lambda\rho H_{4_i}^T \; \sqrt{2}\lambda C_{1_i}^T \end{bmatrix}^T,$$

$$\tilde{D}_{12} = \begin{bmatrix} 0 \; 0 \; 0 \; -\sqrt{2}\lambda I \end{bmatrix}^T, \quad \tilde{D}_{21_i} = \begin{bmatrix} 0 \; 0 \; \delta I \; D_{21_i} \; I \end{bmatrix}$$

$$\textit{and } \lambda = \left(1 + \rho^2 \sum_{i=1}^{r} \sum_{j=1}^{r} \left[\| H_{2_i}^T H_{2_j} \| + \| H_{7_i}^T H_{7_j} \| \right] \right)^{\frac{1}{2}},$$

then the prescribed $\mathcal{H}_\infty$ performance $\gamma > 0$ is guaranteed. Furthermore, a suitable filter is of the form (4.4) with

$$\begin{aligned} \hat{A}_{ij} &= \left[Y^{-1} - X \right]^{-1} \mathcal{M}_{ij} Y^{-1} \\ \hat{B}_i &= \left[Y^{-1} - X \right]^{-1} \mathcal{B}_i \\ \hat{C}_i &= \mathcal{C}_i Y^{-1} \end{aligned} \tag{4.19}$$

where

$$\begin{aligned} \mathcal{M}_{ij} = &-A_i^T - X A_i Y - \gamma^{-2} \left\{ X \tilde{B}_{1_i} + \left[Y^{-1} - X \right] \hat{B}_i \tilde{D}_{21_i}(\mu) \right\} \tilde{B}_{1_j}^T \\ &- \left[Y^{-1} - X \right] \hat{B}_i C_{2_j} Y - \tilde{C}_{1_i}^T \left[\tilde{C}_{1_j} Y + \tilde{D}_{12_j} \hat{C}_j Y \right]. \end{aligned} \tag{4.20}$$

Proof: Suppose there exist X and Y such that the inequalities (4.10) and (4.11)-(4.12) hold. The inequality (4.10) implies that the matrix P defined in (4.7) is a positive definite matrix. Using the partition (4.8), the filter (4.9) and multiplying (4.7) to the left by $\begin{bmatrix} Y & I \\ Y & 0 \end{bmatrix}$ and to the right by $\begin{bmatrix} Y & Y \\ I & 0 \end{bmatrix}$, we have

$$\begin{bmatrix} \Phi_{11_{ij}} & 0 \\ 0 & \Phi_{22_{ij}} \end{bmatrix} \tag{4.21}$$

where

$$\begin{aligned} \Phi_{11_{ij}} &= A_i Y + Y A_i^T + \gamma^{-2} \tilde{B}_{1_i} \tilde{B}_{1_j}^T + \left[Y \tilde{C}_{1_i}^T + \mathcal{C}_i^T \tilde{D}_{12_j}^T \right] \left[Y \tilde{C}_{1_i}^T + \mathcal{C}_i^T \tilde{D}_{12_j}^T \right]^T \\ \Phi_{22_{ij}} &= A_i^T X + X A_i + \mathcal{B}_i C_{2_j} + C_{2_i}^T \mathcal{B}_j^T + \tilde{C}_{1_i}^T \tilde{C}_{1_j} \\ &\quad + \gamma^{-2} \left[X \tilde{B}_{1_i} + \mathcal{B}_i \tilde{D}_{21_j} \right] \left[X \tilde{B}_{1_i} + \mathcal{B}_i \tilde{D}_{21_j} \right]^T. \end{aligned}$$

Note that $\Phi_{11_{ij}}$ and $\Phi_{22_{ij}}$ are the Schur complements of $\Psi_{11_{ij}}$ and $\Psi_{22_{ij}}$. Using (4.13)-(4.16), we have (4.21) less than zero. Hence, by Theorem 4, we learn that the inequality (4.3) holds. ∎

4.1.2 Case II–$\nu(t)$ is unavailable for feedback

Now, the premise variable of the fuzzy model $\nu(t)$ is unavailable for feedback, which implies μ_i is unavailable for feedback. Hence, we cannot select our filter which depends on μ_i. Thus, we select our filter as follows [91]:

$$\begin{aligned} \dot{\hat{x}}(t) &= \sum_{i=1}^{r} \sum_{j=1}^{r} \hat{\mu}_i \hat{\mu}_j \left[\hat{A}_{ij} \hat{x}(t) + \hat{B}_i y(t) \right] \\ \hat{z}(t) &= \sum_{i=1}^{r} \hat{\mu}_i \hat{C}_i \hat{x}(t) \end{aligned} \tag{4.22}$$

where $\hat{\mu}_i$ depends on the premise variable of the filter which is different from μ_i. Figure 4.3 shows the block diagram of the robust $\mathcal{H}_\infty$ filtering problem associated with uncertain fuzzy system in case that μ_i is unavailable for feedback.

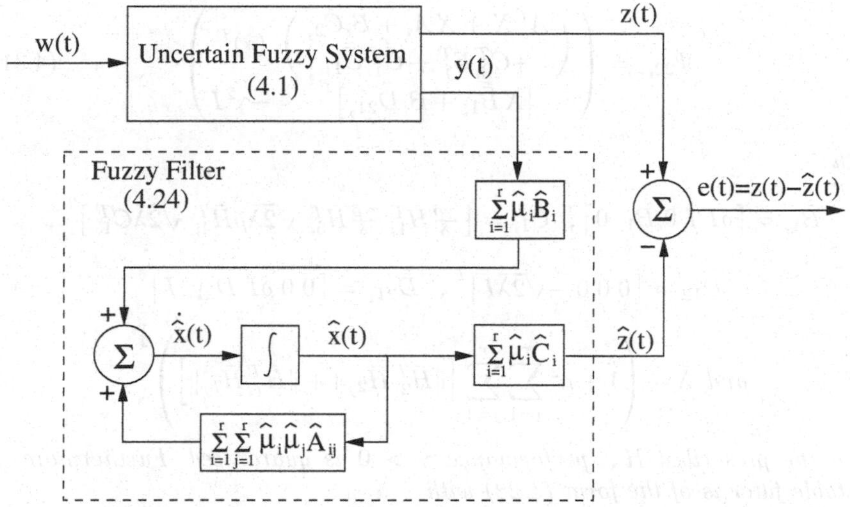

Fig. 4.3. Block diagram of an uncertain fuzzy system with a robust $\mathcal{H}_\infty$ fuzzy filter in Case II.

By applying the same technique used in Subsection 3.3.2, we have the following theorem.

Theorem 5. *Consider the system (4.1). Given a prescribed $\mathcal{H}_\infty$ performance $\gamma > 0$ and a positive constant δ, if there exist matrices $X = X^T$, $Y = Y^T$, $\mathcal{B}_i$ and $\mathcal{C}_i$, $i = 1, 2, \cdots, r$, satisfying the following linear matrix inequalities:*

$$\begin{bmatrix} X & I \\ I & Y \end{bmatrix} > 0 \tag{4.23}$$

$$X > 0 \tag{4.24}$$

$$Y > 0 \tag{4.25}$$

$$\Psi 11_{ii} < 0, \quad i = 1, 2, \cdots, r \tag{4.26}$$

$$\Psi 22_{ii} < 0, \quad i = 1, 2, \cdots, r \tag{4.27}$$

$$\Psi 11_{ij} + \Psi 11_{ji} < 0, \quad i < j \leq r \tag{4.28}$$

$$\Psi 22_{ij} + \Psi 22_{ji} < 0, \quad i < j \leq r \tag{4.29}$$

where

$$\Psi 11_{ij} = \begin{pmatrix} A_i Y + Y A_i^T + \gamma^{-2} \tilde{\bar{B}}_{1_i} \tilde{\bar{B}}_{1_j}^T & (*)^T \\ \left[Y \tilde{\bar{C}}_{1_i}^T + \mathcal{C}_i^T \tilde{\bar{D}}_{12}^T \right]^T & -I \end{pmatrix} \tag{4.30}$$

$$\Psi 22_{ij} = \begin{pmatrix} \begin{pmatrix} A_i^T X + X A_i + \mathcal{B}_i C_{2_j} \\ + C_{2_i}^T \mathcal{B}_j^T + \tilde{\bar{C}}_{1_i}^T \tilde{\bar{C}}_{1_j} \end{pmatrix} & (*)^T \\ \left[X \tilde{\bar{B}}_{1_i} + \mathcal{B}_i \tilde{\bar{D}}_{21_j} \right]^T & -\gamma^2 I \end{pmatrix} \tag{4.31}$$

with

$$\tilde{\bar{B}}_{1_i} = \left[\delta I \; I \; 0 \; B_{1_i} \; 0 \right], \quad \tilde{\bar{C}}_{1_i} = \left[\tfrac{\gamma \bar{\rho}}{\delta} \bar{H}_{1_i}^T \; \tfrac{\gamma \bar{\rho}}{\delta} \bar{H}_{5_i}^T \; \sqrt{2} \bar{\lambda} \bar{\rho} \bar{H}_{4_i}^T \; \sqrt{2} \bar{\lambda} C_{1_i}^T \right]^T,$$

$$\tilde{\bar{D}}_{12} = \left[0 \; 0 \; 0 \; -\sqrt{2} \bar{\lambda} I \right]^T, \quad \tilde{\bar{D}}_{21_i} = \left[0 \; 0 \; \delta I \; D_{21_i} \; I \right]$$

$$\text{and } \bar{\lambda} = \left(1 + \bar{\rho}^2 \sum_{i=1}^{r} \sum_{j=1}^{r} \left[\| \bar{H}_{2_i}^T \bar{H}_{2_j} \| + \| \bar{H}_{7_i}^T \bar{H}_{7_j} \| \right] \right)^{\frac{1}{2}},$$

then the prescribed $\mathcal{H}_\infty$ performance $\gamma > 0$ is guaranteed. Furthermore, a suitable filter is of the form (4.22) with

$$\begin{aligned} \hat{A}_{ij} &= \left[Y^{-1} - X \right]^{-1} \mathcal{M}_{ij} Y^{-1} \\ \hat{B}_i &= \left[Y^{-1} - X \right]^{-1} \mathcal{B}_i \\ \hat{C}_i &= \mathcal{C}_i Y^{-1} \end{aligned} \tag{4.32}$$

where

$$\begin{aligned} \mathcal{M}_{ij} = & -A_i^T - X A_i Y - \left[Y^{-1} - X \right] \hat{B}_i C_{2_j} Y - \tilde{\bar{C}}_{1_i}^T \left[\tilde{\bar{C}}_{1_j} Y + \tilde{\bar{D}}_{12} \hat{C}_j Y \right] \\ & - \gamma^{-2} \left\{ X \tilde{\bar{B}}_{1_i} + \left[Y^{-1} - X \right] \hat{B}_i \tilde{\bar{D}}_{21_i} \right\} \tilde{\bar{B}}_{1_j}^T. \end{aligned} \tag{4.33}$$

Proof: It can be shown by employing the same technique used in the proof for Theorem 3. ∎

4.2 Example

Consider a tunnel diode circuit shown in Figure 4.4 where the tunnel diode is characterized by

$$i_D(t) = 0.002v_D(t) + 0.01v_D^3(t).$$

Let $x_1(t) = v_C(t)$ and $x_2(t) = i_L(t)$ as the state variables, then the circuit is

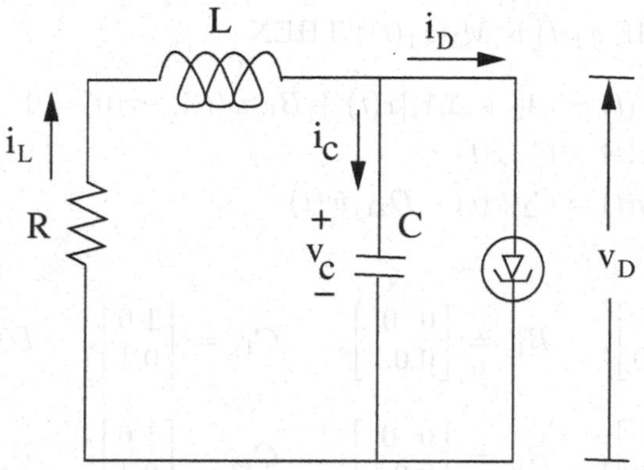

Fig. 4.4. Tunnel diode circuit.

governed by the following state equations:

$$\begin{aligned}
C\dot{x}_1(t) &= -0.002x_1(t) - 0.01x_1^3(t) + x_2(t) \\
L\dot{x}_2(t) &= -x_1(t) - Rx_2(t) + 0.1w_2(t) \\
y(t) &= Jx(t) + 0.1w_1(t) \\
z(t) &= \left[x_1^T(t)\ x_2^T(t)\right]^T
\end{aligned} \tag{4.34}$$

where $w(t)$ is the disturbance noise input, $y(t)$ is the measurement output, $z(t)$ is the state to be estimated and J is the sensor matrix. Note that the variables $x_1(t)$ and $x_2(t)$ are treated as the deviation variables (variables deviate from the desired trajectories). The parameters in the circuit are given as follows: $C = 20\ mF$, $L = 1000\ mH$ and $R = 10 \pm 10\%\ \Omega$. With these parameters, (4.34) can be rewritten as

$$\begin{aligned}
\dot{x}_1(t) &= -0.1x_1(t) - (0.5x_1^2(t)) \cdot x_1(t) + 50x_2(t) \\
\dot{x}_2(t) &= -x_1(t) - (10 + \Delta R)x_2(t) + 0.1w_2(t) \\
y(t) &= Jx(t) + 0.1w_1(t) \\
z(t) &= \left[x_1^T(t)\ x_2^T(t)\right]^T .
\end{aligned} \tag{4.35}$$

For the sake of simplicity, we will use as few rules as possible. Assuming that $|x_1(t)| \leq 3$, the nonlinear network system (4.35) can be approximated by the following TS fuzzy model:

Plant Rule 1: IF $x_1(t)$ is $M_1(x_1(t))$ THEN

$$\begin{aligned} \dot{x}(t) &= [A_1 + \Delta A_1]x(t) + B_{1_1}w(t), \quad x(0) = 0, \\ z(t) &= C_{1_1}x(t), \\ y(t) &= C_{2_1}x(t) + D_{21_1}w(t). \end{aligned}$$

Plant Rule 2: IF $x_1(t)$ is $M_2(x_1(t))$ THEN

$$\begin{aligned} \dot{x}(t) &= [A_2 + \Delta A_2]x(t) + B_{1_2}w(t), \quad x(0) = 0, \\ z(t) &= C_{1_2}x(t), \\ y(t) &= C_{2_2}x(t) + D_{21_2}w(t) \end{aligned}$$

where

$$A_1 = \begin{bmatrix} -0.1 & 50 \\ -1 & -10 \end{bmatrix}, \quad B_{1_1} = \begin{bmatrix} 0 & 0 \\ 0 & 0.1 \end{bmatrix}, \quad C_{1_1} = \begin{bmatrix} 1 & 0 \\ 0 & 1 \end{bmatrix}, \quad D_{21_1} = \begin{bmatrix} 0.1 & 0 \end{bmatrix},$$

$$A_2 = \begin{bmatrix} -4.6 & 50 \\ -1 & -10 \end{bmatrix}, \quad B_{1_2} = \begin{bmatrix} 0 & 0 \\ 0 & 0.1 \end{bmatrix}, \quad C_{1_2} = \begin{bmatrix} 1 & 0 \\ 0 & 1 \end{bmatrix}, \quad D_{21_2} = \begin{bmatrix} 0.1 & 0 \end{bmatrix},$$

$$C_{2_1} = C_{2_2} = J, \quad \Delta A_1 = F(x(t), t)H_{1_1} \text{ and } \Delta A_2 = F(x(t), t)H_{1_2}.$$

Now, by assuming that $\|F(x(t), t)\| \leq \rho = 1$ and since the values of R are uncertain but bounded within 10% of their nominal values given in (4.34), we have

$$H_{1_1} = H_{1_2} = \begin{bmatrix} 0 & 0 \\ 0 & 1 \end{bmatrix}.$$

Figure 4.5 shows the plots of the membership functions for Rules 1 and 2.

Case I-$\nu(t)$ is available for feedback

In this case, $x_1(t) = \nu(t)$ is assumed to be available for feedback; for instance, $J = [1 \ \ 0]$. This implies that μ_i is available for feedback. Using the LMI optimization algorithm and Theorem 4 with $\gamma = 1$ and $\delta = 1$, we obtain the following results:

$$X = \begin{bmatrix} 34.5536 & -2.4910 \\ -2.4910 & 0.8883 \end{bmatrix}, \qquad Y = \begin{bmatrix} 0.8986 & 1.7528 \\ 1.7528 & 27.4284 \end{bmatrix},$$

$$\hat{A}_{11} = \begin{bmatrix} -9.4003 & -1.1377 \\ 63.2915 & -3.7526 \end{bmatrix}, \qquad \hat{A}_{12} = \begin{bmatrix} -14.7653 & -1.3877 \\ 79.5268 & -2.9644 \end{bmatrix},$$

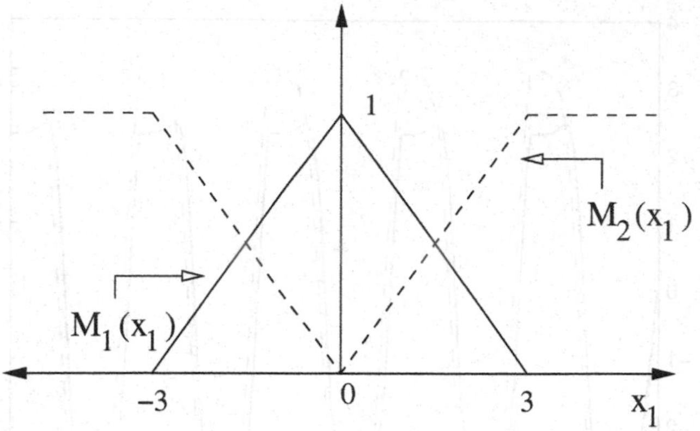

Fig. 4.5. Membership functions for the two fuzzy set.

$$\hat{A}_{21} = \begin{bmatrix} -5.8973 & -0.9794 \\ 49.9964 & -4.1557 \end{bmatrix}, \qquad \hat{A}_{22} = \begin{bmatrix} -11.3243 & -1.2277 \\ 61.4584 & -3.3736 \end{bmatrix},$$

$$\hat{B}_1 = \begin{bmatrix} -0.1292 \\ 1.0828 \end{bmatrix}, \qquad \hat{B}_2 = \begin{bmatrix} -0.0312 \\ 0.7246 \end{bmatrix},$$

$$\hat{C}_1 = \begin{bmatrix} -35.3890 & -1.5720 \end{bmatrix}, \qquad \hat{C}_2 = \begin{bmatrix} -34.7556 & -1.5891 \end{bmatrix}.$$

Hence, the resulting fuzzy filter is

$$\dot{\hat{x}}(t) = \sum_{i=1}^{2}\sum_{j=1}^{2} \mu_i \mu_j \hat{A}_{ij}\hat{x}(t) + \sum_{i=1}^{2} \mu_i \hat{B}_i y(t)$$

$$\hat{z}(t) = \sum_{i=1}^{2} \mu_i \hat{C}_i \hat{x}(t)$$

where

$$\mu_1 = M_1(x_1(t)) \text{ and } \mu_2 = M_2(x_1(t)).$$

Case II-$\nu(t)$ is unavailable for feedback

In this case, $x_1(t) = \nu(t)$ is assumed to be unavailable for feedback; for instance, $J = [0 \;\; 1]$. This implies that μ_i is unavailable for feedback. Using the LMI optimization algorithm and Theorem 5 with $\gamma = 1$ and $\delta = 1$, we obtain the following results:

$$X = \begin{bmatrix} 77.3789 & -15.9191 \\ -15.9191 & 5.3505 \end{bmatrix}, \qquad Y = \begin{bmatrix} 1.6782 & -1.6006 \\ -1.6006 & 27.7695 \end{bmatrix},$$

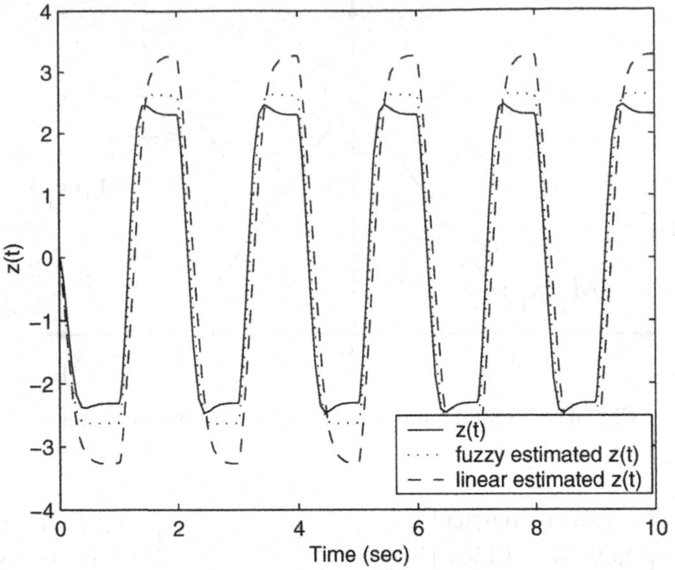

(a) Case I: The histories of $z(t)$

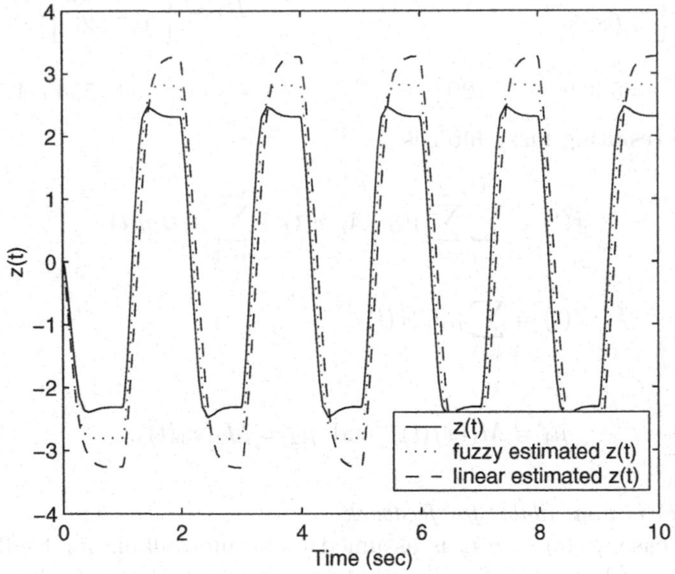

(b) Case II: The histories of $z(t)$

Fig. 4.6. The histories of $z(t)$ in Cases I and II.

$$\hat{A}_{11} = \begin{bmatrix} -11.2662 & -1.0266 \\ 92.4376 & -1.1388 \end{bmatrix}, \qquad \hat{A}_{12} = \begin{bmatrix} -15.0307 & -0.9550 \\ 132.5941 & -2.1872 \end{bmatrix},$$

$$\hat{A}_{21} = \begin{bmatrix} -11.8923 & -1.0235 \\ 77.7075 & -0.9892 \end{bmatrix}, \qquad \hat{A}_{22} = \begin{bmatrix} -15.6465 & -0.9299 \\ 115.8735 & -1.7290 \end{bmatrix},$$

$$\hat{B}_1 = \begin{bmatrix} -0.0181 \\ 0.5905 \end{bmatrix}, \qquad \hat{B}_2 = \begin{bmatrix} -0.0261 \\ 0.3875 \end{bmatrix},$$

$$\hat{C}_1 = \begin{bmatrix} -111.7427 & 2.4315 \end{bmatrix}, \qquad \hat{C}_2 = \begin{bmatrix} -111.7360 & 2.4290 \end{bmatrix}.$$

Hence, the resulting fuzzy filter is

$$\dot{\hat{x}}(t) = \sum_{i=1}^{2}\sum_{j=1}^{2} \hat{\mu}_i \hat{\mu}_j \hat{A}_{ij}\hat{x}(t) + \sum_{i=1}^{2} \hat{\mu}_i \hat{B}_i y(t)$$

$$\hat{z}(t) = \sum_{i=1}^{2} \hat{\mu}_i \hat{C}_i \hat{x}(t)$$

where

$$\hat{\mu}_1 = M_1(\hat{x}_1(t)) \text{ and } \hat{\mu}_2 = M_2(\hat{x}_1(t)).$$

Remark 2. Figure 4.6 shows the responses of $z(t)$. The disturbance input signal, $w(t)$, which was used during the simulation is given in Figure 4.7. The

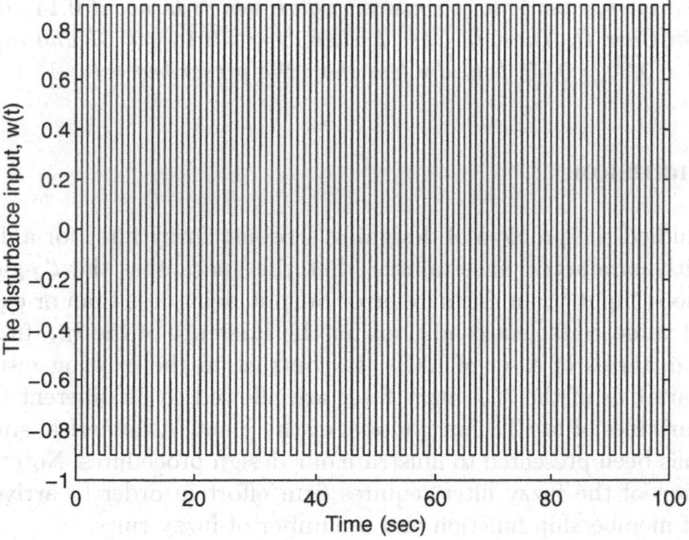

Fig. 4.7. The disturbance input noise, $w(t)$.

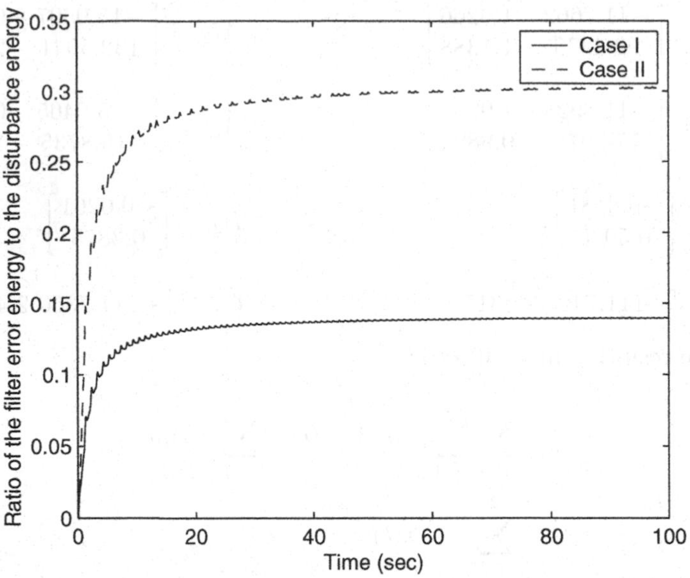

Fig. 4.8. The ratio of the filter error energy to the disturbance noise energy: $\left(\frac{\int_0^{T_f}(z(t)-\hat{z}(t))^T(z(t)-\hat{z}(t))dt}{\int_0^{T_f} w^T(t)w(t)dt}\right)$.

simulation results for the ratio of the filter error energy to the disturbance input noise energy obtained by using the $\mathcal{H}_\infty$ fuzzy filter are depicted in Figure 4.8. After 100 *seconds, the ratio of the filter error energy to the disturbance input noise energy tends to a constant value which is about* 0.14 *in Case I and* 0.30 *in Case II. Thus, in Case I where* $\gamma = \sqrt{0.14} = 0.37$ *and in Case II where* $\gamma = \sqrt{0.30} = 0.55$*, both are less than the prescribed value* 1. □

4.3 Conclusion

We has studied the problem of designing a robust fuzzy filter for a TS fuzzy system with parametric uncertainties that the guarantees the $\mathcal{L}_2$-gain from an exogenous input to an estimate error output being less than or equal to a prescribed value. Sufficient conditions for the existence of the $\mathcal{H}_\infty$ fuzzy filter are given in terms of a set of LMIs. In contrast to the existing results, the premise variables of the $\mathcal{H}_\infty$ fuzzy filter are allowed to be different from the premise variables of the TS fuzzy model of the plant. A numerical simulation example has been presented to illustrate our design procedures. Note that the optimization of the fuzzy filter requires firm effort in order to arrive at the optimal of membership function and a number of fuzzy rules.

5

Robust $\mathcal{H}_\infty$ Fuzzy Control Design for Uncertain Fuzzy Markovian Jump Systems

This chapter deals with the problem of designing a robust fuzzy state and output feedback controller for a TS fuzzy system with MJs and parametric uncertainties. We develop a technique for designing a robust fuzzy controller that guarantees the $\mathcal{L}_2$-gain of the mapping from the exogenous input noise to the regulated output being less than the prescribed value. Sufficient conditions for the existence of a robust $\mathcal{H}_\infty$ fuzzy controller have been derived by solving a set of linear matrix inequalities.

5.1 System Description

We further generalize the TS fuzzy system with parametric uncertainties to a TS fuzzy system with MJs and parametric uncertainties as follows:

$$\begin{aligned}\dot{x}(t) &= \sum_{i=1}^{r} \mu_i(\nu(t))\Big[[A_i(\eta(t)) + \Delta A_i(\eta(t))]x(t)+\\ &\qquad [B_{1_i}(\eta(t)) + \Delta B_{1_i}(\eta(t))]w(t) + [B_{2_i}(\eta(t)) + \Delta B_{2_i}(\eta(t))]u(t)\Big],\\ z(t) &= \sum_{i=1}^{r} \mu_i(\nu(t))\Big[[C_{1_i}(\eta(t)) + \Delta C_{1_i}(\eta(t))]x(t)\\ &\qquad +[D_{12_i}(\eta(t)) + \Delta D_{12_i}(\eta(t))]u(t)\Big]\\ y(t) &= \sum_{i=1}^{r} \mu_i(\nu(t))\Big[[C_{2_i}(\eta(t)) + \Delta C_{2_i}(\eta(t))]x(t)\\ &\qquad +[D_{21_i}(\eta(t)) + \Delta D_{21_i}(\eta(t))]w(t)\Big]\end{aligned} \tag{5.1}$$

with $x(0) = 0$, where $\nu(t) = [\nu_1(t) \ \cdots \ \nu_\vartheta(t)]$ is the premise variable vector that may depend on states in many cases, $\mu_i(\nu(t))$ denotes the normalized time-varying fuzzy weighting functions for each rule (i.e., $\mu_i(\nu(t)) \geq 0$ and $\sum_{i=1}^{r} \mu_i(\nu(t)) = 1$), ϑ is the number of fuzzy sets, $x(t) \in \Re^n$ is the state vector, $u(t) \in \Re^m$ is the input, $w(t) \in \Re^p$ is the disturbance which belongs to $\mathcal{L}_2[0,\infty)$, $y(t) \in \Re^\ell$ is the measurement, $z(t) \in \Re^s$ is the controlled output, and the matrix functions $A_i(\eta(t))$, $B_{1_i}(\eta(t))$,

$B_{2_i}(\eta(t))$, $C_{1_i}(\eta(t))$, $C_{2_i}(\eta(t))$, $D_{12_i}(\eta(t))$, $D_{21_i}(\eta(t))$, $\Delta A_i(\eta(t))$, $\Delta B_{1_i}(\eta(t))$, $\Delta B_{2_i}(\eta(t))$, $\Delta C_{1_i}(\eta(t))$, $\Delta C_{2_i}(\eta(t))$, $\Delta D_{12_i}(\eta(t))$ and $\Delta D_{21_i}(\eta(t))$ are of appropriate dimensions. $\{\eta(t))\}$ is a continuous-time discrete-state Markov process taking values in a finite set $\mathcal{S} = \{1, 2, \cdots, s\}$ with transition probability matrix $Pr \stackrel{\Delta}{=} \{P_{\imath k}(t)\}$ given by

$$
\begin{aligned}
P_{\imath k}(t) &= Pr(\eta(t+\Delta) = k|\eta(t) = \imath) \\
&= \left\{ \begin{array}{ll} \lambda_{\imath k}\Delta + O(\Delta) & \text{if } \imath \neq k \\ 1 + \lambda_{\imath\imath}\Delta + O(\Delta) & \text{if } \imath = k \end{array} \right\}
\end{aligned}
\tag{5.2}
$$

where $\Delta > 0$, and $\lim_{\Delta \longrightarrow 0} \frac{O(\Delta)}{\Delta} = 0$. Here $\lambda_{\imath k} \geq 0$ is the transition rate from mode $\imath$ (system operating mode) to mode k ($\imath \neq k$), and

$$
\lambda_{\imath\imath} = - \sum_{k=1, k\neq \imath}^{s} \lambda_{\imath k}. \tag{5.3}
$$

For the convenience of notations, we let $\mu_i \stackrel{\Delta}{=} \mu_i(\nu(t))$, $\eta = \eta(t)$, and any matrix $M(\mu, \imath) \stackrel{\Delta}{=} M(\mu, \eta = \imath)$. The matrix functions $\Delta A_i(\eta)$, $\Delta B_{1_i}(\eta)$, $\Delta B_{2_i}(\eta)$, $\Delta C_{1_i}(\eta)$, $\Delta C_{2_i}(\eta)$, $\Delta D_{12_i}(\eta)$ and $\Delta D_{21_i}(\eta)$ represent the time-varying uncertainties in the system and satisfy the following assumption.

Assumption 5.1

$$
\begin{aligned}
&\Delta A_i(\eta) = F(x(t), \eta, t)H_{1_i}(\eta),\ \Delta B_{1_i}(\eta) = F(x(t), \eta, t)H_{2_i}(\eta), \\
&\Delta B_{2_i}(\eta) = F(x(t), \eta, t)H_{3_i}(\eta),\ \Delta C_{1_i}(\eta) = F(x(t), \eta, t)H_{4_i}(\eta), \\
&\Delta C_{2_i}(\eta) = F(x(t), \eta, t)H_{5_i}(\eta),\ \Delta D_{12_i}(\eta) = F(x(t), \eta, t)H_{6_i}(\eta), \\
&\textit{and } \Delta D_{21_i}(\eta) = F(x(t), \eta, t)H_{7_i}(\eta)
\end{aligned}
$$

where $H_{j_i}(\eta)$, $j = 1, 2, \cdots, 7$ are known matrices which characterize the structure of the uncertainties. Furthermore, there exists a positive function $\rho(\eta)$ such that the following inequality holds:

$$
\|F(x(t), \eta, t)\| \leq \rho(\eta). \tag{5.4}
$$

Definition 2. *Suppose γ is a given positive real number. A system of the form (5.1) is said to have an $\mathcal{L}_2$-gain less than or equal to γ if*

$$
\mathbf{E}\left[\int_0^{T_f} \left\{ z^T(t)z(t) - \gamma^2 w^T(t)w(t) \right\} dt \right] < 0,\ x(0) = 0 \tag{5.5}
$$

where $\mathbf{E}[\cdot]$ denotes as the expectation operator, for all $T_f \geq 0$ and $w(t) \in [0, T_f]$.

5.2 Robust $\mathcal{H}_\infty$ State-Feedback Control Design

The aim of this section is to design a robust $\mathcal{H}_\infty$ fuzzy state-feedback controller of the form

$$u(t) = \sum_{j=1}^{r} \mu_j K_j(\eta) x(t) \tag{5.6}$$

where $K_j(\eta)$ is the controller gain, such that the inequality (5.5) is guaranteed. The state space form of the fuzzy system model (5.1) with the controller (5.6) is given by

$$\begin{aligned}\dot{x}(t) = \textstyle\sum_{i=1}^{r}\sum_{j=1}^{r} \mu_i\mu_j \Big[&[(A_i(\imath) + B_{2_i}(\imath)K_j(\imath)) + (\Delta A_i(\imath) + \Delta B_{2_i}(\imath)K_j(\imath))]\times \\ &x(t) + [B_{1_i}(\imath) + \Delta B_{1_i}(\imath)]w(t)\Big], \quad x(0) = 0.\end{aligned} \tag{5.7}$$

The following theorem provides sufficient conditions for the existence of a robust $\mathcal{H}_\infty$ fuzzy state-feedback controller. These sufficient conditions can be derived by the Lyapunov approach.

Theorem 6. *Consider the system (5.1). Given a prescribed $\mathcal{H}_\infty$ performance $\gamma > 0$, then the inequality (5.5) holds if for $\imath = 1, 2, \cdots, s$, there exist matrices $P(\imath) = P^T(\imath)$ and any positive constants $\delta(\imath)$ such that the following linear matrix inequalities hold:*

$$P(\imath) > 0 \tag{5.8}$$

$$\Psi_{ii}(\imath) < 0, \quad i = 1, 2, \cdots, r \tag{5.9}$$

$$\Psi_{ij}(\imath) + \Psi_{ji}(\imath) < 0, \quad i < j \le r \tag{5.10}$$

where

$$\Psi_{ij}(\imath) = \begin{pmatrix} \Phi_{ij}(\imath) & (*)^T & (*)^T & (*)^T \\ \mathcal{R}(\imath)\tilde{B}_{1_i}^T(\imath) & -\gamma\mathcal{R}(\imath) & (*)^T & (*)^T \\ \Upsilon_{ij}(\imath) & 0 & -\gamma\mathcal{R}(\imath) & (*)^T \\ \mathcal{Z}^T(\imath) & 0 & 0 & -\mathcal{P}(\imath) \end{pmatrix} \tag{5.11}$$

$$\Phi_{ij}(\imath) = A_i(\imath)P(\imath) + P(\imath)A_i^T(\imath) + B_{2_i}(\imath)Y_j(\imath) + Y_j^T(\imath)B_{2_i}^T(\imath) + \lambda_{\imath\imath}P(\imath) \tag{5.12}$$

$$\Upsilon_{ij}(\imath) = \tilde{C}_{1_i}(\imath)P(\imath) + \tilde{D}_{12_i}(\imath)Y_j(\imath) \tag{5.13}$$

$$\mathcal{R}(\imath) = diag\{\delta(\imath)I, I, \delta(\imath)I, I\} \tag{5.14}$$

$$\mathcal{Z}(\imath) = \left(\sqrt{\lambda_{\imath 1}}P(\imath) \;\cdots\; \sqrt{\lambda_{\imath(\imath-1)}}P(\imath) \;\sqrt{\lambda_{\imath(\imath+1)}}P(\imath) \;\cdots\; \sqrt{\lambda_{\imath s}}P(\imath)\right) \tag{5.15}$$

$$\mathcal{P}(\imath) = diag\{P(1), \cdots, P(\imath-1), P(\imath+1), \cdots, P(s)\} \tag{5.16}$$

with

$$\tilde{B}_{1_i}(\imath) = \begin{bmatrix} I \; I \; I \; B_{1_i}(\imath) \end{bmatrix} \tag{5.17}$$

$$\tilde{C}_{1_i}(\imath) = \begin{bmatrix} \gamma\rho(\imath)H_{1_i}^T(\imath) \; \sqrt{2}\aleph(\imath)\rho(\imath)H_{4_i}^T(\imath) \; 0 \; \sqrt{2}\aleph(\imath)C_{1_i}^T(\imath) \end{bmatrix}^T \tag{5.18}$$

$$\tilde{D}_{12_i}(\imath) = \begin{bmatrix} 0 \; \sqrt{2}\aleph(\imath)\rho(\imath)H_{6_i}^T(\imath) \; \gamma\rho(\imath)H_{3_i}^T(\imath) \; \sqrt{2}\aleph(\imath)D_{12_i}^T(\imath) \end{bmatrix}^T \tag{5.19}$$

$$\aleph(\imath) = \left(1 + \rho^2(\imath)\sum_{i=1}^{r}\sum_{j=1}^{r}\left[\|H_{2_i}^T(\imath)H_{2_j}(\imath)\|\right]\right)^{\frac{1}{2}}. \tag{5.20}$$

Furthermore, a suitable choice of the fuzzy controller is

$$u(t) = \sum_{j=1}^{r}\mu_j K_j(\imath)x(t) \tag{5.21}$$

where

$$K_j(\imath) = Y_j(\imath)(P(\imath))^{-1}. \tag{5.22}$$

Proof: See Appendix. ■

5.3 Robust $\mathcal{H}_\infty$ Output Feedback Control Design

This section aims at designing a full order dynamic $\mathcal{H}_\infty$ fuzzy output feedback controller of the form

$$\begin{aligned} \dot{\hat{x}}(t) &= \textstyle\sum_{i=1}^{r}\sum_{j=1}^{r}\hat{\mu}_i\hat{\mu}_j\left[\hat{A}_{ij}(\imath)\hat{x}(t) + \hat{B}_i(\imath)y(t)\right] \\ u(t) &= \textstyle\sum_{i=1}^{r}\hat{\mu}_i\hat{C}_i(\imath)\hat{x}(t) \end{aligned} \tag{5.23}$$

where $\hat{x}(t) \in \Re^n$ is the controller's state vector, $\hat{A}_{ij}(\imath)$, $\hat{B}_i(\imath)$ and $\hat{C}_i(\imath)$ are parameters of the controller which are to be determined, and $\hat{\mu}_i$ denotes the normalized time-varying fuzzy weighting functions for each rule (i.e., $\hat{\mu}_i \geq 0$ and $\sum_{i=1}^{r}\hat{\mu}_i = 1$), such that the inequality (5.5) holds. Clearly, in real control problems, all of the premise variables are not necessarily measurable. Thus, we can consider the designing of the robust $\mathcal{H}_\infty$ output feedback control into two cases as follows. In Subsection 5.3.1, we consider the case where the premise variable of the fuzzy model μ_i is measurable, while in Subsection 5.3.2, the premise variable which is assumed to be unmeasurable is considered.

5.3.1 Case I–$\nu(t)$ is available for feedback

In this subsection, the premise variable of the fuzzy model $\nu(t)$ is assumed to be available for feedback. This enables us to design a controller which is μ_i-dependent, i.e.,

$$\begin{aligned} \dot{\hat{x}}(t) &= \textstyle\sum_{i=1}^{r}\sum_{j=1}^{r}\mu_i\mu_j\left[\hat{A}_{ij}(\imath)\hat{x}(t) + \hat{B}_i(\imath)y(t)\right] \\ u(t) &= \textstyle\sum_{i=1}^{r}\mu_i\hat{C}_i(\imath)\hat{x}(t). \end{aligned} \tag{5.24}$$

Before presenting our next results, the following lemma is recalled.

Lemma 3. *Consider the system (5.1). Given a prescribed $\mathcal{H}_\infty$ performance $\gamma > 0$ and any positive constants $\delta(\imath)$, for $\imath = 1, 2, \cdots, s$, if there exists a matrix function $P(\imath) = P^T(\imath)$ satisfying the following linear matrix inequalities:*

$$P(\imath) > 0 \quad (5.25)$$

$$\begin{pmatrix} P(\imath)A_{cl}^{ij}(\imath) + (A_{cl}^{ij}(\imath))^T P(\imath) + \sum_{k=1}^{s} \lambda_{\imath k} P(k) & (*)^T & (*)^T \\ (P(\imath)B_{cl}^{ij}(\imath))^T & -\gamma^2 I & (*)^T \\ C_{cl}^{ij}(\imath) & 0 & -I \end{pmatrix} < 0 \quad (5.26)$$

where $i, j = 1, 2, \cdots, r$,

$$A_{cl}^{ij}(\imath) = \begin{bmatrix} A_i(\imath) & B_{2_i}(\imath)\hat{C}_j(\imath) \\ \hat{B}_i(\imath)C_{2_j}(\imath) & \hat{A}_{ij}(\imath) \end{bmatrix},$$

$$B_{cl}^{ij}(\imath) = \begin{bmatrix} \tilde{B}_{1_i}(\imath) \\ \hat{B}_i(\imath)\tilde{D}_{21_j}(\imath) \end{bmatrix} \quad \text{and} \quad C_{cl}^{ij}(\imath) = [\tilde{C}_{1_i}(\imath) \;\; \tilde{D}_{12_i}(\imath)\hat{C}_j(\imath)]$$

with

$$\tilde{B}_{1_i}(\imath) = [\delta(\imath)I \;\; I \;\; \delta(\imath)I \;\; 0 \;\; B_{1_i}(\imath) \;\; 0]$$

$$\tilde{C}_{1_i}(\imath) = \left[\tfrac{\gamma\rho(\imath)}{\delta(\imath)} H_{1_i}^T(\imath) \;\; 0 \;\; \tfrac{\gamma\rho(\imath)}{\delta(\imath)} H_{5_i}^T(\imath) \;\; \sqrt{2}\aleph(\imath)\rho(\imath)H_{4_i}^T(\imath) \;\; \sqrt{2}\aleph(\imath)C_{1_i}^T(\imath) \right]^T$$

$$\tilde{D}_{12_i}(\imath) = \left[0 \;\; \tfrac{\gamma\rho(\imath)}{\delta(\imath)} H_{3_i}^T(\imath) \;\; 0 \;\; \sqrt{2}\aleph(\imath)\rho(\imath)H_{6_i}^T(\imath) \;\; \sqrt{2}\aleph(\imath)D_{12_i}^T(\imath) \right]^T$$

$$\tilde{D}_{21_i}(\imath) = [0 \;\; 0 \;\; 0 \;\; \delta(\imath)I \;\; D_{21_i}(\imath) \;\; I]$$

$$\aleph(\imath) = \left(1 + \rho^2(\imath) \sum_{i=1}^{r}\sum_{j=1}^{r} \left[\|H_{2_i}^T(\imath)H_{2_j}(\imath)\| + \|H_{7_i}^T(\imath)H_{7_j}(\imath)\| \right] \right)^{\frac{1}{2}},$$

then the inequality (5.5) is guaranteed.

Proof: See Appendix. ∎

Knowing that the controller's premise variable is the same as the plant's premise variable, the left hand side of (5.26) can be re-expressed as follows:

$$\begin{aligned} &P(\imath)A_{cl}^{ij}(\imath) + (A_{cl}^{ij}(\imath))^T P(\imath) + \gamma^{-2}P(\imath)B_{cl}^{ij}(\imath)(B_{cl}^{ij}(\imath))^T P(\imath) \\ &+ \textstyle\sum_{k=1}^{s} \lambda_{\imath k} P(k) + (C_{cl}^{ij}(\imath))^T C_{cl}^{ij}(\imath). \end{aligned} \quad (5.27)$$

Before providing LMI-based sufficient conditions for the system (5.1) to have the $\mathcal{H}_\infty$ performance, let us partition the matrix $P(\imath)$ as follows:

$$P(\imath) = \begin{bmatrix} X(\imath) & Y^{-1}(\imath) - X(\imath) \\ Y^{-1}(\imath) - X(\imath) & X(\imath) - Y^{-1}(\imath) \end{bmatrix} \quad (5.28)$$

where $X(\imath) = X^T(\imath) \in \Re^{n\times n}$ and $Y(\imath) = Y^T(\imath) \in \Re^{n\times n}$. Utilizing the partition above, we define the new controller's matrices as

$$\begin{aligned} \mathcal{B}(\mu,\imath) &\triangleq \left[Y^{-1}(\imath) - X(\imath)\right]\hat{B}(\mu,\imath) \\ \mathcal{C}(\mu,\imath) &\triangleq \hat{C}(\mu,\imath)Y(\imath). \end{aligned} \tag{5.29}$$

The following theorem provides LMI-based sufficient conditions for the system (5.1) to have an $\mathcal{H}_\infty$ performance γ with μ_i being available for feedback.

Theorem 7. *Consider the system (5.1). Given a prescribed $\mathcal{H}_\infty$ performance $\gamma > 0$ and any positive constants $\delta(\imath)$, for $\imath = 1, 2, \cdots, s$, if there exist matrices $X(\imath) = X^T(\imath)$, $Y(\imath) = Y^T(\imath)$, $\mathcal{B}_i(\imath)$ and $\mathcal{C}_i(\imath)$, $i = 1, 2, \cdots, r$, satisfying the following linear matrix inequalities:*

$$\begin{bmatrix} X(\imath) & I \\ I & Y(\imath) \end{bmatrix} > 0 \tag{5.30}$$

$$X(\imath) > 0 \tag{5.31}$$

$$Y(\imath) > 0 \tag{5.32}$$

$$\Psi_{11_{ii}}(\imath) < 0, \quad i = 1, 2, \cdots, r \tag{5.33}$$

$$\Psi_{22_{ii}}(\imath) < 0, \quad i = 1, 2, \cdots, r \tag{5.34}$$

$$\Psi_{11_{ij}}(\imath) + \Psi_{11_{ji}}(\imath) < 0, \quad i < j \le r \tag{5.35}$$

$$\Psi_{22_{ij}}(\imath) + \Psi_{22_{ji}}(\imath) < 0, \quad i < j \le r \tag{5.36}$$

where

$$\Psi_{11_{ij}}(\imath) = \begin{pmatrix} \begin{pmatrix} A_i(\imath)Y(\imath) + Y(\imath)A_i^T(\imath) \\ +\lambda_{\imath\imath}Y(\imath) + \gamma^{-2}\tilde{B}_{1_i}(\imath)\tilde{B}_{1_j}^T(\imath) \\ +B_{2_i}(\imath)\mathcal{C}_j(\imath) + \mathcal{C}_i^T(\imath)B_{2_j}^T(\imath) \end{pmatrix} & (*)^T & (*)^T \\ \tilde{C}_{1_i}(\imath)Y(\imath) + \tilde{D}_{12_i}(\imath)\mathcal{C}_j(\imath) & -I & (*)^T \\ \mathcal{J}^T(\imath) & 0 & -\mathcal{Y}(\imath) \end{pmatrix} \tag{5.37}$$

$$\Psi_{22_{ij}}(\imath) = \begin{pmatrix} \begin{pmatrix} A_i^T(\imath)X(\imath) + X(\imath)A_i(\imath) \\ +\mathcal{B}_i(\imath)C_{2_j}(\imath) + C_{2_i}^T(\imath)\mathcal{B}_j^T(\imath) \\ +\tilde{C}_{1_i}^T(\imath)\tilde{C}_{1_j}(\imath) + \sum_{k=1}^{s}\lambda_{\imath k}X(k) \end{pmatrix} & (*)^T \\ \left[X(\imath)\tilde{B}_{1_i}(\imath) + \mathcal{B}_i(\imath)\tilde{D}_{21_j}(\imath)\right]^T & -\gamma^2 I \end{pmatrix} \tag{5.38}$$

with

$$\mathcal{J}(\imath) = \left[\sqrt{\lambda_{1\imath}}Y(\imath) \ \cdots \ \sqrt{\lambda_{(i-1)\imath}}Y(\imath) \ \sqrt{\lambda_{(i+1)\imath}}Y(\imath) \ \cdots \ \sqrt{\lambda_{s\imath}}Y(\imath)\right]$$

$$\mathcal{Y}(\imath) = diag\left\{Y(1), \ \cdots, \ Y(\imath-1), \ Y(\imath+1), \ \cdots \ , Y(s)\right\}$$

$$\tilde{B}_{1_i}(\imath) = [\delta(\imath)I \ I \ \delta(\imath)I \ 0 \ B_{1_i}(\imath) \ 0]$$

$$\tilde{C}_{1_i}(\imath) = \left[\tfrac{\gamma\rho(\imath)}{\delta(\imath)}H_{1_i}^T(\imath) \ 0 \ \tfrac{\gamma\rho(\imath)}{\delta(\imath)}H_{5_i}^T(\imath) \ \sqrt{2}\aleph(\imath)\rho(\imath)H_{4_i}^T(\imath) \ \sqrt{2}\aleph(\imath)C_{1_i}^T(\imath)\right]^T$$

$$\tilde{D}_{12_i}(\imath) = \left[0 \ \tfrac{\gamma\rho(\imath)}{\delta(\imath)}H_{3_i}^T(\imath) \ 0 \ \sqrt{2}\aleph(\imath)\rho(\imath)H_{6_i}^T(\imath) \ \sqrt{2}\aleph(\imath)D_{12_i}^T(\imath)\right]^T$$

$$\tilde{D}_{21_i}(\imath) = [0 \ 0 \ 0 \ \delta(\imath)I \ D_{21_i}(\imath) \ I]$$

$$\aleph(\imath) = \left(1+\rho^2(\imath)\sum_{i=1}^{r}\sum_{j=1}^{r}\left[\|H_{2_i}^T(\imath)H_{2_j}(\imath)\| + \|H_{7_i}^T(\imath)H_{7_j}(\imath)\|\right]\right)^{\frac{1}{2}},$$

then the prescribed $\mathcal{H}_\infty$ performance $\gamma > 0$ is guaranteed. Furthermore, a suitable controller is of the form (5.24) with

$$\begin{aligned} \hat{A}_{ij}(\imath) &= \left[Y^{-1}(\imath) - X(\imath)\right]^{-1}\mathcal{M}_{ij}(\imath)Y^{-1}(\imath) \\ \hat{B}_i(\imath) &= \left[Y^{-1}(\imath) - X(\imath)\right]^{-1}\mathcal{B}_i(\imath) \\ \hat{C}_i(\imath) &= \mathcal{C}_i(\imath)Y^{-1}(\imath) \end{aligned} \tag{5.39}$$

where

$$\begin{aligned} \mathcal{M}_{ij}(\imath) = &-A_i^T(\imath) - X(\imath)A_i(\imath)Y(\imath) - \left[Y^{-1}(\imath) - X(\imath)\right]\hat{B}_i(\imath)C_{2_j}(\imath)Y(\imath) \\ &-X(\imath)B_{2_i}(\imath)\hat{C}_j(\imath)Y(\imath) - \textstyle\sum_{k=1}^{s}\lambda_{\imath k}Y^{-1}(k)Y(\imath) \\ &-\tilde{C}_{1_i}^T(\imath)\left[\tilde{C}_{1_j}(\imath)Y(\imath) + \tilde{D}_{12_j}(\imath)\hat{C}_j(\imath)Y(\imath)\right] \\ &-\gamma^{-2}\left\{X(\imath)\tilde{B}_{1_i}(\imath) + \left[Y^{-1}(\imath) - X(\imath)\right]\hat{B}_i(\imath)\tilde{D}_{21_i}(\imath)\right\}\tilde{B}_{1_j}^T(\imath). \end{aligned} \tag{5.40}$$

Proof: Suppose there exist $X(\imath)$ and $Y(\imath)$ such that the inequalities (5.30) and (5.31)-(5.32) hold. The inequality (5.30) implies that the matrix P defined in (5.27) is a positive definite matrix. Using the partition (5.28), the controller (5.29) and multiplying (5.27) to the left by $\begin{bmatrix} Y(\imath) & I \\ Y(\imath) & 0 \end{bmatrix}$ and to the right by $\begin{bmatrix} Y(\imath) & Y(\imath) \\ I & 0 \end{bmatrix}$, we have

$$\begin{bmatrix} \Phi_{11_{ij}}(\imath) & 0 \\ 0 & \Phi_{22_{ij}}(\imath) \end{bmatrix} \tag{5.41}$$

where

$$
\begin{aligned}
\Phi_{11_{ij}}(\imath) &= A_i(\imath)Y(\imath) + Y(\imath)A_i^T(\imath) + B_{2_i}(\imath)\mathcal{C}_j(\imath) + \mathcal{C}_i^T(\imath)B_{2_j}^T(\imath) + \lambda_{\imath\imath}Y(\imath) \\
&\quad + \left[Y(\imath)\tilde{C}_{1_i}^T(\imath) + \mathcal{C}_i^T(\imath)\tilde{D}_{12_j}^T(\imath)\right]\left[Y(\imath)\tilde{C}_{1_i}^T(\imath) + \mathcal{C}_i^T(\imath)\tilde{D}_{12_j}^T(\imath)\right]^T \\
&\quad + \gamma^{-2}\tilde{B}_{1_i}(\imath)\tilde{B}_{1_j}^T(\imath) + \mathcal{J}^T(\imath)\mathcal{Y}^{-1}(\imath)\mathcal{J}(\imath) \qquad (5.42)\\
\Phi_{22_{ij}}(\imath) &= A_i^T(\imath)X(\imath) + X(\imath)A_i(\imath) + \mathcal{B}_i(\imath)C_{2_j}(\imath) + C_{2_i}^T(\imath)\mathcal{B}_j^T(\imath) \\
&\quad + \gamma^{-2}\left[X(\imath)\tilde{B}_{1_i}(\imath) + \mathcal{B}_i(\imath)\tilde{D}_{21_j}(\imath)\right]\left[X(\imath)\tilde{B}_{1_i}(\imath) + \mathcal{B}_i(\imath)\tilde{D}_{21_j}(\imath)\right]^T \\
&\quad + \tilde{C}_{1_i}^T(\imath)\tilde{C}_{1_j}(\imath) + \sum_{k=1}^{s}\lambda_{\imath k}X(k). \qquad (5.43)
\end{aligned}
$$

Note that $\Phi_{11_{ij}}(\imath)$ and $\Phi_{22_{ij}}(\imath)$ are the Schur complements of $\Psi_{11_{ij}}(\imath)$ and $\Psi_{22_{ij}}(\imath)$. Using (5.33)–(5.36), we have (5.41) less than zero. Hence, by Theorem 7, we learn that the inequality (5.5) holds. ■

5.3.2 Case II–$\nu(t)$ is unavailable for feedback

In this case, the fuzzy model's premise variable, $\nu(t)$, is assumed to be unavailable for feedback. Under this assumption, the controller's premise variable cannot be selected to be the same as the plant's premise variable, i.e,

$$
\begin{aligned}
\dot{\hat{x}}(t) &= \sum_{i=1}^{r}\sum_{j=1}^{r}\hat{\mu}_i\hat{\mu}_j\left[\hat{A}_{ij}(\imath)\hat{x}(t) + \hat{B}_i(\imath)y(t)\right] \\
u(t) &= \sum_{i=1}^{r}\hat{\mu}_i\hat{C}_i(\imath)\hat{x}(t)
\end{aligned} \qquad (5.44)
$$

where $\hat{\mu}_i$ is a function of the controller's premise variable which is different from the plant's premise variable.

Now, let us re-express the system (5.1) in terms of $\hat{\mu}_i$, thus the plant's premise variable becomes the same as the controller's premise variable. By doing so, the result given in the previous case can then be applied here. Note that it can be done by using the same technique as in Subsection 3.3.2. After some manipulation, we get

$$
\begin{aligned}
\dot{x}(t) &= \sum_{i=1}^{r}\hat{\mu}_i\Big[[A_i(\imath) + \Delta\bar{A}_i(\imath)]x(t) + [B_{1_i}(\imath) + \Delta\bar{B}_{1_i}(\imath)]w(t) \\
&\quad + [B_{2_i}(\imath) + \Delta\bar{B}_{2_i}(\imath)]u(t)\Big], \quad x(0) = 0 \\
z(t) &= \sum_{i=1}^{r}\hat{\mu}_i\Big[[C_{1_i}(\imath) + \Delta\bar{C}_{1_i}(\imath)]x(t) + [D_{12_i}(\imath) + \Delta\bar{D}_{12_i}(\imath)]u(t)\Big] \\
y(t) &= \sum_{i=1}^{r}\hat{\mu}_i\Big[[C_{2_i}(\imath) + \Delta\bar{C}_{2_i}(\imath)]x(t) + [D_{21_i}(\imath) + \Delta\bar{D}_{21_i}(\imath)]w(t)\Big]
\end{aligned} \qquad (5.45)
$$

where

$$\Delta\bar{A}_i(\imath) = \bar{F}(x(t), \hat{x}(t), \imath, t)\bar{H}_{1_i}(\imath), \quad \Delta\bar{B}_{1_i}(\imath) = \bar{F}(x(t), \hat{x}(t), \imath, t)\bar{H}_{2_i}(\imath),$$

$$\Delta\bar{B}_{2_i}(\imath) = \bar{F}(x(t), \hat{x}(t), \imath, t)\bar{H}_{3_i}(\imath), \quad \Delta\bar{C}_{1_i}(\imath) = \bar{F}(x(t), \hat{x}(t), \imath, t)\bar{H}_{4_i}(\imath),$$

$$\Delta\bar{C}_{2_i}(\imath) = \bar{F}(x(t), \hat{x}(t), \imath, t)\bar{H}_{5_i}(\imath), \quad \Delta\bar{D}_{12_i}(\imath) = \bar{F}(x(t), \hat{x}(t), \imath, t)\bar{H}_{6_i}(\imath)$$

and $\Delta\bar{D}_{21_i}(\imath) = \bar{F}(x(t), \hat{x}(t), \imath, t)\bar{H}_{7_i}(\imath)$

with

$$\bar{H}_{1_i}(\imath) = \left[H_{1_i}^T(\imath) \; A_1^T(\imath) \; \cdots \; A_r^T(\imath) \; H_{1_1}^T(\imath) \; \cdots \; H_{1_r}^T(\imath)\right]^T,$$

$$\bar{H}_{2_i}(\imath) = \left[H_{2_i}^T(\imath) \; B_{1_1}^T(\imath) \; \cdots \; B_{1_r}^T(\imath) \; H_{2_1}^T(\imath) \; \cdots \; H_{2_r}^T(\imath)\right]^T,$$

$$\bar{H}_{3_i}(\imath) = \left[H_{3_i}^T(\imath) \; B_{2_1}^T(\imath) \; \cdots \; B_{2_r}^T(\imath) \; H_{3_1}^T(\imath) \; \cdots \; H_{3_r}^T(\imath)\right]^T,$$

$$\bar{H}_{4_i}(\imath) = \left[H_{4_i}^T(\imath) \; C_{1_1}^T(\imath) \; \cdots \; C_{1_r}^T(\imath) \; H_{4_1}^T(\imath) \; \cdots \; H_{4_r}^T(\imath)\right]^T,$$

$$\bar{H}_{5_i}(\imath) = \left[H_{5_i}^T(\imath) \; C_{2_1}^T(\imath) \; \cdots \; C_{2_r}^T(\imath) \; H_{5_1}^T(\imath) \; \cdots \; H_{5_r}^T(\imath)\right]^T,$$

$$\bar{H}_{6_i}(\imath) = \left[H_{6_i}^T(\imath) \; D_{12_1}^T(\imath) \; \cdots \; D_{12_r}^T(\imath) \; H_{6_1}^T(\imath) \; \cdots \; H_{6_r}^T(\imath)\right]^T,$$

$$\bar{H}_{7_i}(\imath) = \left[H_{7_i}^T(\imath) \; D_{21_1}^T(\imath) \; \cdots \; D_{21_r}^T(\imath) \; H_{7_1}^T(\imath) \; \cdots \; H_{7_r}^T(\imath)\right]^T$$

and $\bar{F}(x(t), \hat{x}(t), \imath, t) = \Big[F(x(t), \imath, t) \; (\mu_1 - \hat{\mu}_1) \; \cdots \; (\mu_r - \hat{\mu}_r) \; F(x(t), \imath, t)(\mu_1 - \hat{\mu}_1) \; \cdots \; F(x(t), \imath, t)(\mu_r - \hat{\mu}_r)\Big]$. Note that $\|\bar{F}(x(t), \hat{x}(t), \imath, t)\| \leq \bar{\rho}(\imath)$ where $\bar{\rho}(\imath) = \{3\rho^2(\imath) + 2\}^{\frac{1}{2}}$. $\bar{\rho}(\imath)$ is derived by utilizing the concept of vector norm in the basic system control theory and the fact that $\mu_i \geq 0$, $\hat{\mu}_i \geq 0$, $\sum_{i=1}^r \mu_i = 1$ and $\sum_{i=1}^r \hat{\mu}_i = 1$.

In this new expression, the plant's premise variable is now the same as the controller's premise variable. Note that the above technique is basically employed in order to obtain the plant's premise variable to be the same as the controller's premise variable; e.g. [27, 28]. Thus, applying Theorem 7, we have the following LMI-based sufficient conditions for this case.

Theorem 8. *Consider the system (5.1). Given a prescribed $\mathcal{H}_\infty$ performance $\gamma > 0$ and any positive constants $\delta(\imath)$, for $\imath = 1, 2, \cdots, s$, if there exist matrices $X(\imath) = X^T(\imath)$, $Y(\imath) = Y^T(\imath)$, $\mathcal{B}_i(\imath)$ and $\mathcal{C}_i(\imath)$, $i = 1, 2, \cdots, r$, satisfying the following linear matrix inequalities:*

$$\begin{bmatrix} X(\imath) & I \\ I & Y(\imath) \end{bmatrix} > 0 \quad (5.46)$$

$$X(\imath) > 0 \quad (5.47)$$

$$Y(\imath) > 0 \quad (5.48)$$

$$\Psi_{11_{ii}}(\imath) < 0, \quad i = 1, 2, \cdots, r \quad (5.49)$$

$$\Psi_{22_{ii}}(\imath) < 0, \quad i = 1, 2, \cdots, r \quad (5.50)$$

$$\Psi_{11_{ij}}(\imath) + \Psi_{11_{ji}}(\imath) < 0, \quad i < j \leq r \quad (5.51)$$

$$\Psi_{22_{ij}}(\imath) + \Psi_{22_{ji}}(\imath) < 0, \quad i < j \leq r \quad (5.52)$$

where

$$\Psi_{11_{ij}}(\imath) = \begin{pmatrix} \begin{pmatrix} A_i(\imath)Y(\imath) + Y(\imath)A_i^T(\imath) \\ +\lambda_{\imath\imath}Y(\imath) + \gamma^{-2}\tilde{\bar{B}}_{1_i}(\imath)\tilde{\bar{B}}_{1_j}^T(\imath) \\ +B_{2_i}(\imath)\mathcal{C}_j(\imath) + \mathcal{C}_i^T(\imath)B_{2_j}^T(\imath) \end{pmatrix} & (*)^T & (*)^T \\ \tilde{\bar{C}}_{1_i}(\imath)Y(\imath) + \tilde{\bar{D}}_{12_i}(\imath)\mathcal{C}_j(\imath) & -I & (*)^T \\ \mathcal{J}^T(\imath) & 0 & -\mathcal{Y}(\imath) \end{pmatrix} \tag{5.53}$$

$$\Psi_{22_{ij}}(\imath) = \begin{pmatrix} \begin{pmatrix} A_i^T(\imath)X(\imath) + X(\imath)A_i(\imath) \\ +\mathcal{B}_i(\imath)C_{2_j}(\imath) + C_{2_i}^T(\imath)\mathcal{B}_j^T(\imath) \\ +\tilde{\bar{C}}_{1_i}^T(\imath)\tilde{\bar{C}}_{1_j}(\imath) + \sum_{k=1}^{s}\lambda_{\imath k}X(k) \end{pmatrix} & (*)^T \\ \left[X(\imath)\tilde{\bar{B}}_{1_i}(\imath) + \mathcal{B}_i(\imath)\tilde{\bar{D}}_{21_j}(\imath)\right]^T & -\gamma^2 I \end{pmatrix} \tag{5.54}$$

with

$$\mathcal{J}(\imath) = \left[\sqrt{\lambda_{1\imath}}Y(\imath) \ \cdots \ \sqrt{\lambda_{(i-1)\imath}}Y(\imath) \ \sqrt{\lambda_{(i+1)\imath}}Y(\imath) \ \cdots \ \sqrt{\lambda_{s\imath}}Y(\imath)\right]$$

$$\mathcal{Y}(\imath) = diag\left\{Y(1), \ \cdots, \ Y(\imath-1), \ Y(\imath+1), \ \cdots \ ,Y(s)\right\}$$

$$\tilde{\bar{B}}_{1_i}(\imath) = [\delta(\imath)I \ \ I \ \ \delta(\imath)I \ \ 0 \ \ B_{1_i}(\imath) \ \ 0]$$

$$\tilde{\bar{C}}_{1_i}(\imath) = \left[\tfrac{\gamma\bar{\rho}(\imath)}{\delta(\imath)}\bar{H}_{1_i}^T(\imath) \ \ 0 \ \ \tfrac{\gamma\bar{\rho}(\imath)}{\delta(\imath)}\bar{H}_{5_i}^T(\imath) \ \ \sqrt{2}\bar{\aleph}(\imath)\bar{\rho}(\imath)\bar{H}_{4_i}^T(\imath) \ \ \sqrt{2}\bar{\aleph}(\imath)C_{1_i}^T(\imath)\right]^T$$

$$\tilde{\bar{D}}_{12_i}(\imath) = \left[0 \ \ \tfrac{\gamma\bar{\rho}(\imath)}{\delta(\imath)}\bar{H}_{3_i}^T(\imath) \ \ 0 \ \ \sqrt{2}\bar{\aleph}(\imath)\bar{\rho}(\imath)\bar{H}_{6_i}^T(\imath) \ \ \sqrt{2}\bar{\aleph}(\imath)D_{12_i}^T(\imath)\right]^T$$

$$\tilde{\bar{D}}_{21_i}(\imath) = [0 \ \ 0 \ \ 0 \ \ \delta(\imath)I \ \ D_{21_i}(\imath) \ \ I],$$

$$\bar{\aleph}(\imath) = \left(1 + \bar{\rho}^2(\imath)\sum_{i=1}^{r}\sum_{j=1}^{r}\left[\|\bar{H}_{2_i}^T(\imath)\bar{H}_{2_j}(\imath)\| + \|\bar{H}_{7_i}^T(\imath)\bar{H}_{7_j}(\imath)\|\right]\right)^{\frac{1}{2}},$$

then the prescribed $\mathcal{H}_\infty$ performance $\gamma > 0$ is guaranteed. Furthermore, a suitable controller is of the form (5.44) with

$$\begin{aligned} \hat{A}_{ij}(\imath) &= \left[Y^{-1}(\imath) - X(\imath)\right]^{-1}\mathcal{M}_{ij}(\imath)Y^{-1}(\imath) \\ \hat{B}_i(\imath) &= \left[Y^{-1}(\imath) - X(\imath)\right]^{-1}\mathcal{B}_i(\imath) \\ \hat{C}_i(\imath) &= \mathcal{C}_i(\imath)Y^{-1}(\imath) \end{aligned} \tag{5.55}$$

where

$$\begin{aligned} \mathcal{M}_{ij}(\imath) = &-A_i^T(\imath) - X(\imath)A_i(\imath)Y(\imath) - \left[Y^{-1}(\imath) - X(\imath)\right]\hat{B}_i(\imath)C_{2_j}(\imath)Y(\imath) \\ &-X(\imath)B_{2_i}(\imath)\hat{C}_j(\imath)Y(\imath) - \textstyle\sum_{k=1}^{s}\lambda_{\imath k}Y^{-1}(k)Y(\imath) \\ &-\tilde{\bar{C}}_{1_i}^T(\imath)\left[\tilde{\bar{C}}_{1_j}(\imath)Y(\imath) + \tilde{\bar{D}}_{12_j}(\imath)\hat{C}_j(\imath)Y(\imath)\right] \\ &-\gamma^{-2}\left\{X(\imath)\tilde{\bar{B}}_{1_i}(\imath) + \left[Y^{-1}(\imath) - X(\imath)\right]\hat{B}_i(\imath)\tilde{\bar{D}}_{21_i}(\imath)\right\}\tilde{\bar{B}}_{1_j}^T(\imath). \end{aligned} \tag{5.56}$$

Proof: Since (5.45) is of the form of (5.1), it can be shown by employing the proof for Theorem 7. ■

5.4 Example

Consider a modified Samuelson multiplier-accelerator economic model based on [92] which is governed by the following difference equations:

$$\begin{aligned} Y(k) &= [C(k) + I(k) + G(k-1)] \\ I(k) &= (\alpha + \Delta\alpha)\,[Y(k-1) - Y(k-2)] \\ C(k) &= (\beta + \Delta\beta)Y^{v}(k-1) \end{aligned} \tag{5.57}$$

where Y is the deviation of the national income from the desired national income, I is the deviation of the induced private investment from the desired induced private investment, C is the deviation of the consumption expenditure from the desired consumption expenditure, G is the deviation of the government expenditure from the desired government expenditure decided at the end of period $(k-1)$ for period k, α is the accelerator coefficient, β is the marginal propensity to consume parameter and v is the consume parameter $(v \geq 1)$. $\Delta\alpha$ and $\Delta\beta$ are the uncertain accelerator coefficient and marginal propensity to consume parameter, respectively. We assume that $|\Delta\alpha| \leq 0.1\alpha$ and $|\Delta\beta| \leq 0.1\beta$.

Eliminating $C(k)$ and $I(k)$ in the above equations, we have

$$\begin{aligned} Y(k) = (\beta + \Delta\beta)Y^{v}(k-1) + (\alpha + \Delta\alpha)Y(k-1) + G(k-1) \\ -(\alpha + \Delta\alpha)Y(k-2). \end{aligned} \tag{5.58}$$

By shifting one step forward and giving $x_1(k) = Y(k-1)$, $x_2(k) = Y(k)$ and $u(k) = G(k)$, (5.58) becomes

$$\begin{bmatrix} x_1(k+1) \\ x_2(k+1) \end{bmatrix} = \begin{bmatrix} x_2(k) \\ \begin{pmatrix} -(\alpha + \Delta\alpha)x_1(k) + (\alpha + \Delta\alpha)x_2(k) \\ +(\beta + \Delta\beta)x_2^{v}(k) \end{pmatrix} \end{bmatrix} + \begin{bmatrix} 0 \\ 1 \end{bmatrix} u(k). \tag{5.59}$$

Converting (5.59) to continuous-time system model, we have

$$\begin{bmatrix} \dot{x}_1(t) \\ \dot{x}_2(t) \end{bmatrix} = \begin{bmatrix} x_2(t) + x_1(t) \\ \begin{pmatrix} -(\alpha + \Delta\alpha)x_1(t) \\ +(\alpha + \Delta\alpha)x_2(t) \\ +(\beta + \Delta\beta)x_2^{v}(t) + x_2(t) \end{pmatrix} \end{bmatrix} + \begin{bmatrix} 0 \\ 1 \end{bmatrix} u(t). \tag{5.60}$$

Based on [92], the general economic situations could be aggregated into three modes as shown in Table 5.1:

The transition probability matrix that relates the three operation modes is given as follows:

$$P_{\imath k} = \begin{bmatrix} 0.67 & 0.17 & 0.16 \\ 0.30 & 0.47 & 0.23 \\ 0.26 & 0.10 & 0.64 \end{bmatrix}.$$

Assuming $v = 2$, (5.60) can be re-expressed as

Table 5.1. Economic Terminology.

Mode $\imath$	Terminology	$\alpha(\imath) \pm \Delta\alpha(\imath)$	$\beta(\imath) \pm \Delta\beta(\imath)$
1	Normal	2.5 ±10%	0.3 ±10%
2	Boom	43.7 ±10%	−0.7 ± 10%
3	Slump	−5.3 ± 10%	0.9 ±10%

$$\begin{aligned}
\begin{bmatrix}\dot{x}_1(t)\\ \dot{x}_2(t)\end{bmatrix} &= \begin{bmatrix}1 & 1\\ -\alpha(\imath) & \alpha(\imath)+\beta(\imath)x_2(t)+1\end{bmatrix}\begin{bmatrix}x_1(t)\\ x_2(t)\end{bmatrix} \\
&\quad + \begin{bmatrix}0 & 0\\ 0.1 & 0\end{bmatrix} w(t) + \begin{bmatrix}0\\ 1\end{bmatrix} u(t) \\
&\quad + \begin{bmatrix}0 & 0\\ -\Delta\alpha(\imath) & \Delta\alpha(\imath)+\Delta\beta(\imath)x_2(t)\end{bmatrix}\begin{bmatrix}x_1(t)\\ x_2(t)\end{bmatrix} \\
z(t) &= \begin{bmatrix}x_1(t)\\ x_2(t)\end{bmatrix} \\
y(t) &= Jx(t) + [0\ 0.1]w(t)
\end{aligned} \tag{5.61}$$

where $x_1(t)$ and $x_2(t)$ are the state vectors, $u(t)$ is the controlled input which represents the deviation of the government expenditure from the desired government expenditure, $w(t)$ is the disturbance input which represents the unexpected behavior of the economy, $z(t)$ is the controlled output, $y(t)$ is the measured output and J is the sensor matrix.

The control objective is to control the state variable $x_2(t)$ for the range $x_2(t) \in [N_1\ N_2]$. For the sake of simplicity, we will use as few rules as possible. Note that Figure 5.1 shows the plot of the membership functions represented by

$$M_1(x_2(t)) = \frac{-x_2(t)+N_2}{N_2-N_1} \quad \text{and} \quad M_2(x_2(t)) = \frac{x_2(t)-N_1}{N_2-N_1}.$$

Knowing that $x_2(t) \in [N_1\ N_2]$, the nonlinear system (5.61) can be approximated by the following TS fuzzy model:

Plant Rule 1: IF $x_2(t)$ is $M_1(x_2(t))$ THEN

$$\begin{aligned}
\dot{x}(t) &= [A_1(\imath) + \Delta A_1(\imath)]x(t) + B_{1_1}(\imath)w(t) + B_{2_1}(\imath)u(t), \quad x(0)=0,\\
z(t) &= C_{1_1}(\imath)x(t),\\
y(t) &= C_{2_1}(\imath)x(t) + D_{21_1}(\imath)w(t).
\end{aligned}$$

Plant Rule 2: IF $x_2(t)$ is $M_2(x_2(t))$ THEN

$$\begin{aligned}
\dot{x}(t) &= [A_2(\imath) + \Delta A_2(\imath)]x(t) + B_{1_2}(\imath)w(t) + B_{2_2}(\imath)u(t), \quad x(0)=0,\\
z(t) &= C_{1_2}(\imath)x(t),\\
y(t) &= C_{2_2}(\imath)x(t) + D_{21_2}(\imath)w(t)
\end{aligned}$$

where $x(t) = [x_1^T(t)\ x_2^T(t)]^T$,

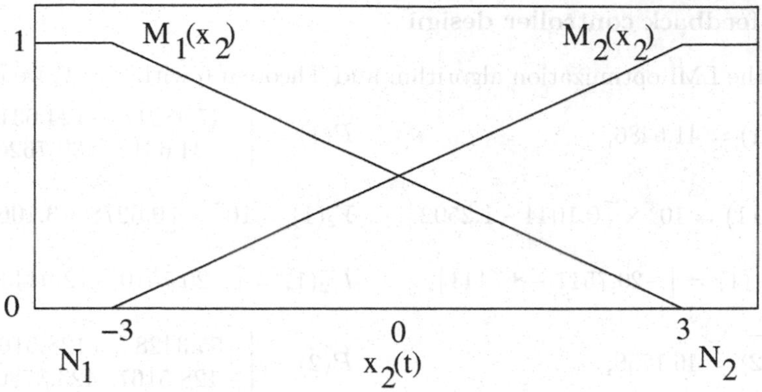

Fig. 5.1. Membership functions for the two fuzzy set.

$$A_1(1) = \begin{bmatrix} 1 & 1 \\ -2.5 & 3.5 + 0.7N_1 \end{bmatrix}, \quad A_2(1) = \begin{bmatrix} 1 & 1 \\ -2.5 & 3.5 + 0.7N_2 \end{bmatrix}$$

$$A_1(2) = \begin{bmatrix} 1 & 1 \\ -43.7 & 44.5 + 1.7N_1 \end{bmatrix}, \quad A_2(2) = \begin{bmatrix} 1 & 1 \\ -43.7 & 44.5 + 1.7N_2 \end{bmatrix}$$

$$A_1(3) = \begin{bmatrix} 1 & 1 \\ 5.3 & -4.3 + 0.1N_1 \end{bmatrix}, \quad A_2(3) = \begin{bmatrix} 1 & 1 \\ 5.3 & -4.3 + 0.1N_2 \end{bmatrix}$$

$$B_{1_1}(1) = B_{1_2}(1) = B_{1_1}(2) = B_{1_2}(2) = B_{1_1}(3) = B_{1_2}(3) = \begin{bmatrix} 0 & 0 \\ 0.1 & 0 \end{bmatrix},$$

$$B_{2_1}(1) = B_{2_2}(1) = B_{2_1}(2) = B_{2_2}(2) = B_{2_1}(3) = B_{2_2}(3) = \begin{bmatrix} 0 \\ 1 \end{bmatrix},$$

$$C_{1_1}(1) = C_{1_2}(1) = C_{1_1}(2) = C_{1_2}(2) = C_{1_1}(3) = C_{1_2}(3) = \begin{bmatrix} 1 & 0 \\ 0 & 1 \end{bmatrix},$$

$$C_{2_1}(1) = C_{2_2}(1) = C_{2_1}(2) = C_{2_2}(2) = C_{2_1}(3) = C_{2_2}(3) = J,$$

$$D_{21_1}(1) = D_{21_2}(1) = D_{21_1}(2) = D_{21_2}(2) = D_{21_1}(3) = D_{21_2}(3) = \begin{bmatrix} 0 & 0.1 \end{bmatrix},$$

$$\Delta A_1(\imath) = F(x(t), \imath, t)H_{1_1}(\imath) \text{ and } \Delta A_2(\imath) = F(x(t), \imath, t)H_{1_2}(\imath).$$

Now, by assuming that in (5.61), $\|F(x(t), \imath, t)\| \le \rho(\imath) = 1$, we have

$$H_{1_1}(\imath) = \begin{bmatrix} 0 & 0 \\ -0.1\alpha(\imath) & 0.1\alpha(\imath) + 0.1\beta(\imath)N_1 \end{bmatrix}$$

$$\text{and } H_{1_2}(\imath) = \begin{bmatrix} 0 & 0 \\ -0.1\alpha(\imath) & 0.1\alpha(\imath) + 0.1\beta(\imath)N_2 \end{bmatrix}.$$

In this simulation, we select $N_1 = -3$ and $N_2 = 3$.

State-feedback controller design

Using the LMI optimization algorithm and Theorem 6 with $\gamma = 1$, we obtain

$$\delta(1) = 41.6486, \qquad P(1) = \begin{bmatrix} 47.4891 & -124.6312 \\ -124.6312 & 439.7626 \end{bmatrix},$$

$$Y_1(1) = 10^3 \times \begin{bmatrix} 0.1044 & -1.2592 \end{bmatrix}, \qquad Y_2(1) = 10^3 \times \begin{bmatrix} 0.6278 & -3.1062 \end{bmatrix},$$

$$K_1(1) = \begin{bmatrix} -20.7511 & -8.7443 \end{bmatrix}, \qquad K_2(1) = \begin{bmatrix} -20.7510 & -12.9443 \end{bmatrix},$$

$$\delta(2) = 46.1759, \qquad P(2) = \begin{bmatrix} 55.3428 & -128.5167 \\ -128.5167 & 423.3780 \end{bmatrix},$$

$$Y_1(2) = 10^4 \times \begin{bmatrix} 0.7315 & -2.2627 \end{bmatrix}, \qquad Y_2(2) = 10^4 \times \begin{bmatrix} 0.8369 & -2.6946 \end{bmatrix},$$

$$K_1(2) = \begin{bmatrix} 27.3538 & -45.1409 \end{bmatrix}, \qquad K_2(2) = \begin{bmatrix} 11.6005 & -60.1228 \end{bmatrix},$$

$$\delta(3) = 43.9140, \qquad P(3) = \begin{bmatrix} 52.1186 & -130.8730 \\ -130.8730 & 448.0091 \end{bmatrix},$$

$$Y_1(3) = 10^3 \times \begin{bmatrix} -1.0664 & 2.4200 \end{bmatrix}, \qquad Y_2(3) = 10^3 \times \begin{bmatrix} -0.9878 & 2.1512 \end{bmatrix},$$

$$K_1(3) = \begin{bmatrix} -25.8817 & -2.1590 \end{bmatrix}, \qquad K_2(3) = \begin{bmatrix} -25.8817 & -2.7590 \end{bmatrix}.$$

The resulting fuzzy controller is

$$u(t) = \sum_{j=1}^{2} \mu_j K_j(\imath) x(t)$$

where

$$\mu_1 = M_1(x_2(t)) \text{ and } \mu_2 = M_2(x_2(t)).$$

Output feedback controller design

Case I: $\nu(t)$ are available for feedback

In this case, $\nu(t) = x_2(t)$ is assumed to be available for feedback; for instance, $J = [0 \quad 1]$. This implies that μ_i is available for feedback. Using the LMI optimization algorithm and Theorem 7 with $\gamma = 1$ and $\delta(\imath) = 1$,we obtain

$$X(1) = \begin{bmatrix} 43.3821 & 69.5718 \\ 69.5718 & 386.9464 \end{bmatrix}, \qquad Y(1) = \begin{bmatrix} 0.4960 & -0.9002 \\ -0.9002 & 3.6918 \end{bmatrix},$$

$$\hat{A}_{11}(1) = \begin{bmatrix} 10.0318 & 1.0079 \\ -123.0107 & -65.2591 \end{bmatrix}, \quad \hat{A}_{12}(1) = \begin{bmatrix} 10.1591 & 0.9003 \\ -118.6456 & -68.7917 \end{bmatrix},$$

$$\hat{A}_{21}(1) = \begin{bmatrix} 10.0079 & 1.0933 \\ -122.9968 & -61.0529 \end{bmatrix}, \quad \hat{A}_{22}(1) = \begin{bmatrix} 10.1352 & 0.9857 \\ -118.6317 & -64.5855 \end{bmatrix},$$

$$\hat{B}_1(1) = \begin{bmatrix} -31.6693 \\ 7.9142 \end{bmatrix}, \quad \hat{B}_2(1) = \begin{bmatrix} -31.5525 \\ 12.0990 \end{bmatrix},$$

$$\hat{C}_1(1) = \begin{bmatrix} -0.5184 & -141.9962 \end{bmatrix}, \quad \hat{C}_2(1) = \begin{bmatrix} -5.8859 & -159.0397 \end{bmatrix},$$

$$X(2) = \begin{bmatrix} 45.5544 & 6.0210 \\ 6.0210 & 13.9210 \end{bmatrix}, \quad Y(2) = \begin{bmatrix} 0.1759 & -0.0224 \\ -0.0224 & 0.1156 \end{bmatrix},$$

$$\hat{A}_{11}(2) = 10^3 \times \begin{bmatrix} 0.1388 & 1.0140 \\ -1.3391 & -9.8999 \end{bmatrix}, \hat{A}_{12}(2) = 10^3 \times \begin{bmatrix} 0.0128 & 0.1141 \\ -0.1237 & -1.1135 \end{bmatrix},$$

$$\hat{A}_{21}(2) = 10^3 \times \begin{bmatrix} 0.0490 & 0.2208 \\ -0.4355 & -3.2840 \end{bmatrix}, \hat{A}_{22}(2) = 10^3 \times \begin{bmatrix} 0.0582 & 0.2511 \\ -0.5624 & -3.7084 \end{bmatrix},$$

$$\hat{B}_1(2) = \begin{bmatrix} -28.3321 \\ 173.4354 \end{bmatrix}, \quad \hat{B}_2(2) = \begin{bmatrix} -30.5197 \\ 204.8014 \end{bmatrix},$$

$$\hat{C}_1(2) = 10^3 \times \begin{bmatrix} -0.1107 & -1.2470 \end{bmatrix}, \quad \hat{C}_2(2) = 10^3 \times \begin{bmatrix} -0.1532 & -1.3909 \end{bmatrix},$$

$$X(3) = \begin{bmatrix} 29.6285 & -39.4287 \\ -39.4287 & 77.8877 \end{bmatrix}, \quad Y(3) = \begin{bmatrix} 0.3094 & -0.3190 \\ -0.3190 & 1.1708 \end{bmatrix},$$

$$\hat{A}_{11}(3) = \begin{bmatrix} -44.6535 & -49.5945 \\ -198.0970 & -193.4395 \end{bmatrix}, \quad \hat{A}_{12}(3) = \begin{bmatrix} -44.7323 & -49.9065 \\ -198.3023 & -194.2173 \end{bmatrix},$$

$$\hat{A}_{21}(3) = \begin{bmatrix} -44.5448 & -48.9764 \\ -198.0298 & -192.4832 \end{bmatrix}, \quad \hat{A}_{22}(3) = \begin{bmatrix} -44.6237 & -49.2884 \\ -198.2351 & -193.2610 \end{bmatrix},$$

$$\hat{B}_1(3) = \begin{bmatrix} 57.1346 \\ 32.6070 \end{bmatrix}, \quad \hat{B}_2(3) = \begin{bmatrix} 57.4491 \\ 33.3825 \end{bmatrix},$$

$$\hat{C}_1(3) = \begin{bmatrix} -173.3989 & -161.6216 \end{bmatrix}, \quad \hat{C}_2(3) = \begin{bmatrix} -173.5604 & -162.2239 \end{bmatrix}.$$

The resulting fuzzy controller is

$$\dot{\hat{x}}(t) = \sum_{i=1}^{2} \sum_{j=1}^{2} \mu_i \mu_j \hat{A}_{ij}(\imath) \hat{x}(t) + \sum_{i=1}^{2} \mu_i \hat{B}_i(\imath) y(t)$$

$$u(t) = \sum_{i=1}^{2} \mu_i \hat{C}_i(\imath) \hat{x}(t)$$

where
$$\mu_1 = M_1(x_2(t)) \text{ and } \mu_2 = M_2(x_2(t)).$$

Case II: $\nu(t)$ are unavailable for feedback

In this case, $x_2(t) = \nu(t)$ is assumed to be unavailable for feedback; for instance, $J = [1\ \ 0]$. This implies that μ_i is unavailable for feedback. Using the LMI optimization algorithm and Theorem 8 with $\gamma = 1$ and $\delta(\imath) = 1$, we obtain

$$X(1) = 10^3 \times \begin{bmatrix} 1.0312 & 0.6497 \\ 0.6497 & 0.5099 \end{bmatrix}, \qquad Y(1) = \begin{bmatrix} 0.5085 & -0.9378 \\ -0.9378 & 3.7992 \end{bmatrix},$$

$$\hat{A}_{11}(1) = \begin{bmatrix} 11.3110 & 3.2504 \\ -386.6777 & -207.3117 \end{bmatrix}, \qquad \hat{A}_{12}(1) = \begin{bmatrix} 11.3178 & 3.2437 \\ -382.1906 & -210.8295 \end{bmatrix},$$

$$\hat{A}_{21}(1) = \begin{bmatrix} 11.0800 & 3.0360 \\ -386.3331 & -202.7875 \end{bmatrix}, \qquad \hat{A}_{22}(1) = \begin{bmatrix} 11.0869 & 3.0293 \\ -381.8459 & -206.3054 \end{bmatrix},$$

$$\hat{B}_1(1) = \begin{bmatrix} -11.3146 \\ 18.0796 \end{bmatrix}, \qquad \hat{B}_2(1) = \begin{bmatrix} -25.0573 \\ 40.0729 \end{bmatrix},$$

$$\hat{C}_1(1) = \begin{bmatrix} -387.2706 & -210.5697 \end{bmatrix}, \qquad \hat{C}_2(1) = \begin{bmatrix} -382.7788 & -214.0931 \end{bmatrix},$$

$$X(2) = \begin{bmatrix} 259.4300 & 32.4118 \\ 32.4118 & 17.3704 \end{bmatrix}, \qquad Y(2) = \begin{bmatrix} 0.1839 & -0.0244 \\ -0.0244 & 0.1146 \end{bmatrix},$$

$$\hat{A}_{11}(2) = 10^3 \times \begin{bmatrix} 0.0453 & 0.3444 \\ -0.4707 & -3.5877 \end{bmatrix}, \qquad \hat{A}_{12}(2) = 10^3 \times \begin{bmatrix} 0.0444 & 0.3581 \\ -0.4609 & -3.7309 \end{bmatrix},$$

$$\hat{A}_{21}(2) = 10^3 \times \begin{bmatrix} 0.0126 & 0.0876 \\ -0.1425 & -1.1073 \end{bmatrix}, \qquad \hat{A}_{22}(2) = 10^3 \times \begin{bmatrix} 0.0136 & 0.0910 \\ -0.1554 & -1.1506 \end{bmatrix},$$

$$\hat{B}_1(2) = \begin{bmatrix} -29.8494 \\ 242.8519 \end{bmatrix}, \qquad \hat{B}_2(2) = \begin{bmatrix} -36.4228 \\ 291.8673 \end{bmatrix},$$

$$\hat{C}_1(2) = 10^3 \times \begin{bmatrix} -0.4706 & -3.9083 \end{bmatrix}, \qquad \hat{C}_2(2) = 10^3 \times \begin{bmatrix} -0.5140 & -4.0561 \end{bmatrix},$$

$$X(3) = \begin{bmatrix} 122.6259 & -36.2691 \\ -36.2691 & 197.3630 \end{bmatrix}, \qquad Y(3) = \begin{bmatrix} 0.3101 & -0.3537 \\ -0.3537 & 1.2375 \end{bmatrix},$$

$$\hat{A}_{11}(3) = \begin{bmatrix} 7.8694 & -1.9796 \\ -591.0791 & -521.5024 \end{bmatrix}, \qquad \hat{A}_{12}(3) = \begin{bmatrix} 7.8668 & -1.9883 \\ -591.2649 & -522.1170 \end{bmatrix},$$

$$\hat{A}_{21}(3) = \begin{bmatrix} 7.8710 & -1.9696 \\ -591.0733 & -520.8918 \end{bmatrix}, \qquad \hat{A}_{22}(3) = \begin{bmatrix} 7.8683 & -1.9782 \\ -591.2591 & -521.5064 \end{bmatrix},$$

$$\hat{B}_1(3) = \begin{bmatrix} 5.5268 \\ 3.1088 \end{bmatrix}, \qquad \hat{B}_2(3) = \begin{bmatrix} 5.4929 \\ 3.3695 \end{bmatrix},$$

$$\hat{C}_1(3) = \begin{bmatrix} -595.5896 & -516.0405 \end{bmatrix}, \quad \hat{C}_2(3) = \begin{bmatrix} -595.7740 & -516.6495 \end{bmatrix}.$$

The resulting fuzzy controller is

$$\dot{\hat{x}}(t) = \sum_{i=1}^{2}\sum_{j=1}^{2} \hat{\mu}_i\hat{\mu}_j\hat{A}_{ij}(\imath)\hat{x}(t) + \sum_{i=1}^{2} \hat{\mu}_i\hat{B}_i(\imath)y(t)$$

$$u(t) = \sum_{i=1}^{2} \hat{\mu}_i\hat{C}_i(\imath)\hat{x}(t)$$

where

$$\hat{\mu}_1 = M_1(\hat{x}_2(t)) \text{ and } \hat{\mu}_2 = M_2(\hat{x}_2(t)).$$

Remark 3. Both robust fuzzy state and output feedback controllers guarantee that the $\mathcal{L}_2$-gain, γ, is less than the prescribed value. Figure 5.2 shows the changing between modes during the simulation with the initial mode 1. The disturbance input signal, $w(t)$, which was used during simulation is the rectangular signal (magnitude 0.1 and frequency 10 Hz). The ratios of the regulated output energy to the disturbance input noise energy for both cases are depicted in Figure 5.3. After time = 3, the ratio of the regulated output energy to

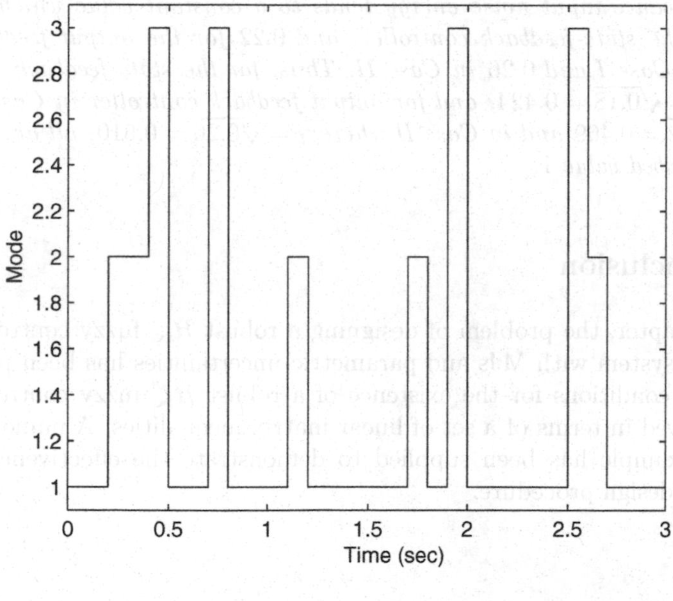

Fig. 5.2. The result of the changing between modes during the simulation with the initial mode 1.

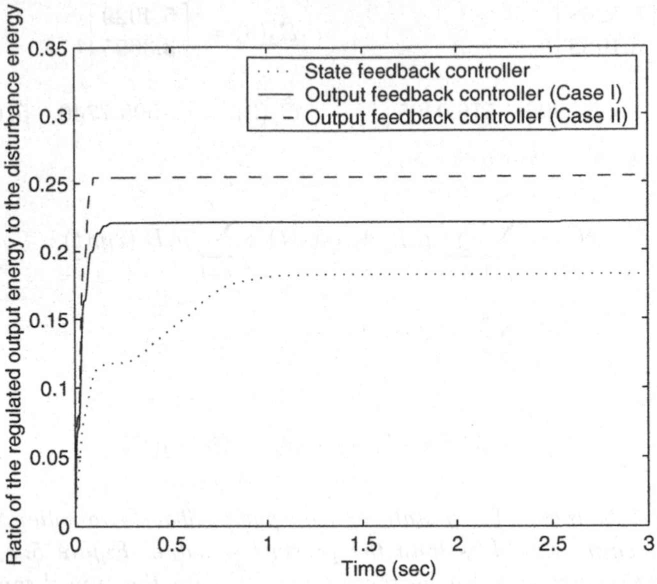

Fig. 5.3. The ratio of the regulated output energy to the disturbance noise energy, $\left(\frac{\int_0^{T_f} z^T(t)z(t)dt}{\int_0^{T_f} w^T(t)w(t)dt}\right)$.

the disturbance input noise energy tends to a constant value which is about 0.18 *for the state-feedback controller, and* 0.22 *for the output feedback controller in Case I and* 0.26 *in Case II. Thus, for the state-feedback controller where* $\gamma = \sqrt{0.18} = 0.424$, *and for output feedback controller in Case I where* $\gamma = \sqrt{0.22} = 0.469$ *and in Case II where* $\gamma = \sqrt{0.26} = 0.510$, *all are less than the prescribed value* 1. □

5.5 Conclusion

In this chapter, the problem of designing a robust $\mathcal{H}_\infty$ fuzzy controller for a TS fuzzy system with MJs and parametric uncertainties has been presented. Sufficient conditions for the existence of a robust $\mathcal{H}_\infty$ fuzzy controller have been derived in terms of a set of linear matrix inequalities. A numerical simulation example has been supplied to demonstrate the effectiveness of the proposed design procedure.

6

Robust $\mathcal{H}_\infty$ Fuzzy Filter Design for Uncertain Fuzzy Markovian Jump Systems

This chapter presents a technique of designing a robust fuzzy filter for a TS fuzzy system with MJs and parametric uncertainties. We develop a technique for designing a robust fuzzy filter such that the $\mathcal{L}_2$-gain of the mapping from the exogenous input noise to the estimated error output is less than the prescribed value. The proposed design is given in terms of LMIs.

6.1 Robust $\mathcal{H}_\infty$ Fuzzy Filter Design

Without loss of generality, we assume $u(t) = 0$. Let us recall the system (5.1) with $u(t) = 0$ as follows:

$$\begin{aligned} \dot{x}(t) &= \sum_{i=1}^{r} \mu_i \Big[[A_i(\eta) + \Delta A_i(\eta)]x(t) + [B_{1_i}(\eta) + \Delta B_{1_i}(\eta)]w(t)\Big], \quad x(0) = 0 \\ z(t) &= \sum_{i=1}^{r} \mu_i \Big[[C_{1_i}(\eta) + \Delta C_{1_i}(\eta)\Big] x(t) \\ y(t) &= \sum_{i=1}^{r} \mu_i \Big[[C_{2_i}(\eta) + \Delta C_{2_i}(\eta)]x(t) + [D_{21_i}(\eta) + \Delta D_{21_i}(\eta)]w(t)\Big]. \end{aligned} \tag{6.1}$$

The aim is to design a full order dynamic $\mathcal{H}_\infty$ fuzzy filter of the form

$$\begin{aligned} \dot{\hat{x}}(t) &= \sum_{i=1}^{r} \sum_{j=1}^{r} \hat{\mu}_i \hat{\mu}_j \Big[\hat{A}_{ij}(\imath)\hat{x}(t) + \hat{B}_i(\imath)y(t)\Big] \\ \hat{z}(t) &= \sum_{i=1}^{r} \hat{\mu}_i \hat{C}_i(\imath)\hat{x}(t) \end{aligned} \tag{6.2}$$

where $\hat{x}(t) \in \Re^n$ is the filter's state vector, $\hat{z} \in \Re^s$ is the estimate of $z(t)$, $\hat{A}_{ij}(\imath)$, $\hat{B}_i(\imath)$ and $\hat{C}_i(\imath)$ are parameters of the filter which are to be determined, and $\hat{\mu}_i$ denotes the normalized time-varying fuzzy weighting functions for each rule (i.e., $\hat{\mu}_i \geq 0$ and $\sum_{i=1}^{r} \hat{\mu}_i = 1$), such that the following inequality holds

$$\mathbf{E}\left[\int_0^{T_f} \left\{\Big(z(t) - \hat{z}(t)\Big)^T \Big(z(t) - \hat{z}(t)\Big) - \gamma^2 w^T(t)w(t)\right\} dt\right] \leq 0, \; x(0) = 0 \tag{6.3}$$

where $\mathbf{E}[\cdot]$ stands for the mathematical expectation and $(z(t) - \hat{z}(t))$ is the estimated error output, for all $T_f \geq 0$ and $w(t) \in [0, T_f]$.

Clearly, in real control problems, all of the premise variables are not necessarily measurable, thus two cases will be considered in this section. Subsection 6.1.1 considers the case where the premise variable of the fuzzy model μ_i is measurable, while in Subsection 6.1.2, the premise variable is assumed to be unmeasurable.

6.1.1 Case I–$\nu(t)$ is available for feedback

The premise variable of the fuzzy model $\nu(t)$ is available for feedback which implies that μ_i is available for feedback. Thus, we can select our filter that depends on μ_i as follows [91]:

$$\begin{aligned}\dot{\hat{x}}(t) &= \textstyle\sum_{i=1}^{r}\sum_{j=1}^{r}\mu_i\mu_j\left[\hat{A}_{ij}(\imath)\hat{x}(t) + \hat{B}_i(\imath)y(t)\right]\\ \hat{z}(t) &= \textstyle\sum_{i=1}^{r}\mu_i\hat{C}_i(\imath)\hat{x}(t).\end{aligned} \tag{6.4}$$

Before presenting our next results, the following lemma is recalled.

Lemma 4. *Consider the system (6.1). Given a prescribed $\mathcal{H}_\infty$ performance $\gamma > 0$ and any positive constants $\delta(\imath)$, for $\imath = 1, 2, \cdots, s$, if there exist matrices $P(\imath) = P^T(\imath)$ such that the following linear inequalities hold:*

$$P(\imath) > 0 \tag{6.5}$$

$$\begin{pmatrix} P(\imath)A_{cl}^{ij}(\imath) + (A_{cl}^{ij}(\imath))^T P(\imath) + \sum_{k=1}^{s}\lambda_{\imath k}P(k) & (*)^T & (*)^T \\ (P(\imath)B_{cl}^{ij}(\imath))^T & -\gamma^2 I & (*)^T \\ C_{cl}^{ij}(\imath) & 0 & -I \end{pmatrix} < 0 \tag{6.6}$$

where $i, j = 1, 2, \cdots, r$,

$$A_{cl}^{ij}(\imath) = \begin{bmatrix} A_i(\imath) & 0 \\ \hat{B}_i(\imath)C_{2_j}(\imath) & \hat{A}_{ij}(\imath) \end{bmatrix},$$

$$B_{cl}^{ij}(\imath) = \begin{bmatrix} \tilde{B}_{1_i}(\imath) \\ \hat{B}_i(\imath)\tilde{D}_{21_j}(\imath) \end{bmatrix}, \quad C_{cl}^{ij}(\imath) = [\tilde{C}_{1_i}(\imath) \quad \tilde{D}_{12}(\imath)\hat{C}_j(\imath)]$$

with

$$\begin{aligned}
\tilde{B}_{1_i}(\imath) &= [\delta(\imath)I \ \ I \ \ 0 \ \ B_{1_i}(\imath) \ \ 0] \\
\tilde{C}_{1_i}(\imath) &= \left[\tfrac{\gamma\rho(\imath)}{\delta(\imath)}H_{1_i}^T(\imath) \ \ \tfrac{\gamma\rho(\imath)}{\delta(\imath)}H_{5_i}^T(\imath) \ \ \sqrt{2}\aleph(\imath)\rho(\imath)H_{4_i}^T(\imath) \ \ \sqrt{2}\aleph(\imath)C_{1_i}^T(\imath)\right]^T \\
\tilde{D}_{12}(\imath) &= \left[0 \ \ 0 \ \ 0 \ \ -\sqrt{2}\aleph(\imath)I\right]^T \\
\tilde{D}_{21_i}(\imath) &= [0 \ \ 0 \ \ \delta(\imath)I \ \ D_{21_i}(\imath) \ \ I]
\end{aligned}$$

$$\aleph(\imath) = \left(1 + \rho^2(\imath)\sum_{i=1}^{r}\sum_{j=1}^{r}\left[\|H_{2_i}^T(\imath)H_{2_j}(\imath)\| + \|H_{7_i}^T(\imath)H_{7_j}(\imath)\|\right]\right)^{\frac{1}{2}},$$

then the inequality (6.3) is guaranteed.

Proof: The proof can be carried out by the same technique used in Lemma 5.1. ∎

Knowing that the filter's premise variable is the same as the plant's premise variable, the left hand side of (6.6) can be re-expressed as follows:

$$P(\imath)A_{cl}^{ij}(\imath) + (A_{cl}^{ij}(\imath))^T P(\imath) + \gamma^{-2}P(\imath)B_{cl}^{ij}(\imath)(B_{cl}^{ij}(\imath))^T P(\imath) + \sum_{k=1}^{s} \lambda_{\imath k} P(k) + (C_{cl}^{ij}(\imath))^T C_{cl}^{ij}(\imath). \tag{6.7}$$

Before providing LMI-based sufficient conditions for the system (5.1) with $u(t) = 0$ to have an $\mathcal{H}_\infty$ performance, let us partition the matrix $P(\imath)$ given by Lemma 4 as follows:

$$P(\imath) = \begin{bmatrix} X(\imath) & Y^{-1}(\imath) - X(\imath) \\ Y^{-1}(\imath) - X(\imath) & X(\imath) - Y^{-1}(\imath) \end{bmatrix} \tag{6.8}$$

where $X(\imath) = X^T(\imath) \in \Re^{n\times n}$ and $Y(\imath) = Y^T(\imath) \in \Re^{n\times n}$. Utilizing the partition above, we define the new filter's input and output matrices as

$$\begin{aligned} \mathcal{B}_i(\imath) &\triangleq \left[Y^{-1}(\imath) - X(\imath)\right]\hat{B}_i(\imath) \\ \mathcal{C}_i(\imath) &\triangleq \hat{C}_i(\imath)Y(\imath). \end{aligned} \tag{6.9}$$

Using these changes of variable, we have the following theorem.

Theorem 9. *Consider the system (6.1). Given a prescribed $\mathcal{H}_\infty$ performance $\gamma > 0$ and any positive constants $\delta(\imath)$, for $\imath = 1, 2, \cdots, s$, if there exist matrices $X(\imath) = X^T(\imath)$, $Y(\imath) = Y^T(\imath)$, $\mathcal{B}_i(\imath)$ and $\mathcal{C}_i(\imath)$, $i = 1, 2, \cdots, r$, satisfying the following linear matrix inequalities:*

$$\begin{bmatrix} X(\imath) & I \\ I & Y(\imath) \end{bmatrix} > 0 \tag{6.10}$$

$$X(\imath) > 0 \tag{6.11}$$

$$Y(\imath) > 0 \tag{6.12}$$

$$\Psi_{11_{ii}}(\imath) < 0, \quad i = 1, 2, \cdots, r \tag{6.13}$$

$$\Psi_{22_{ii}}(\imath) < 0, \quad i = 1, 2, \cdots, r \tag{6.14}$$

$$\Psi_{11_{ij}}(\imath) + \Psi_{11_{ji}}(\imath) < 0, \quad i < j \le r \tag{6.15}$$

$$\Psi_{22_{ij}}(\imath) + \Psi_{22_{ji}}(\imath) < 0, \quad i < j \le r \tag{6.16}$$

where

$$\Psi_{11_{ij}}(\imath) = \begin{pmatrix} \begin{pmatrix} A_i(\imath)Y(\imath) + Y(\imath)A_i^T(\imath) \\ +\lambda_{\imath\imath}Y(\imath) + \gamma^{-2}\tilde{B}_{1_i}(\imath)\tilde{B}_{1_j}^T(\imath) \end{pmatrix} & (*)^T & (*)^T \\ \tilde{C}_{1_i}(\imath)Y(\imath) + \tilde{D}_{12}(\imath)\mathcal{C}_j(\imath) & -I & (*)^T \\ \mathcal{J}^T(\imath) & 0 & -\mathcal{Y}(\imath) \end{pmatrix} \tag{6.17}$$

$$\Psi_{22_{ij}}(\imath) = \begin{pmatrix} \begin{pmatrix} A_i^T(\imath)X(\imath) + X(\imath)A_i(\imath) \\ +\mathcal{B}_i(\imath)C_{2_j}(\imath) + C_{2_i}^T(\imath)\mathcal{B}_j^T(\imath) \\ +\tilde{C}_{1_i}^T(\imath)\tilde{C}_{1_j}(\imath) + \sum_{k=1}^s \lambda_{\imath k}X(k) \end{pmatrix} & (*)^T \\ \tilde{B}_{1_i}^T(\imath)X(\imath) + \tilde{D}_{21_i}^T(\imath)\mathcal{B}_j^T(\imath) & -\gamma^2 I \end{pmatrix} \tag{6.18}$$

with

$$\begin{aligned}
\mathcal{J}(\imath) &= \left[\sqrt{\lambda_{1\imath}}Y(\imath) \ \cdots \ \sqrt{\lambda_{(i-1)\imath}}Y(\imath) \ \sqrt{\lambda_{(i+1)\imath}}Y(\imath) \ \cdots \ \sqrt{\lambda_{s\imath}}Y(\imath)\right] \\
\mathcal{Y}(\imath) &= diag\left\{Y(1), \ \cdots, \ Y(\imath-1), \ Y(\imath+1), \ \cdots \ ,Y(s)\right\} \\
\tilde{B}_{1_i}(\imath) &= [\delta(\imath)I \ I \ 0 \ B_{1_i}(\imath) \ 0] \\
\tilde{C}_{1_i}(\imath) &= \left[\tfrac{\gamma\rho(\imath)}{\delta(\imath)}H_{1_i}^T(\imath) \ \tfrac{\gamma\rho(\imath)}{\delta(\imath)}H_{5_i}^T(\imath) \ \sqrt{2}\aleph(\imath)\rho(\imath)H_{4_i}^T(\imath) \ \sqrt{2}\aleph(\imath)C_{1_i}^T(\imath)\right]^T \\
\tilde{D}_{12}(\imath) &= \left[0 \ 0 \ 0 \ -\sqrt{2}\aleph(\imath)I\right]^T \\
\tilde{D}_{21_i}(\imath) &= [0 \ 0 \ \delta(\imath)I \ D_{21_i}(\imath) \ I] \\
\aleph(\imath) &= \left(1 + \rho^2(\imath)\sum_{i=1}^r\sum_{j=1}^r\left[\|H_{2_i}^T(\imath)H_{2_j}(\imath)\| + \|H_{7_i}^T(\imath)H_{7_j}(\imath)\|\right]\right)^{\frac{1}{2}},
\end{aligned}$$

then the prescribed $\mathcal{H}_\infty$ performance $\gamma > 0$ is guaranteed. Furthermore, a suitable filter is of the form (6.4) with

$$\begin{aligned}
\hat{A}_{ij}(\imath) &= \left[Y^{-1}(\imath) - X(\imath)\right]^{-1}\mathcal{M}_{ij}(\imath)Y^{-1}(\imath) \\
\hat{B}_i(\imath) &= \left[Y^{-1}(\imath) - X(\imath)\right]^{-1}\mathcal{B}_i(\imath) \\
\hat{C}_i(\imath) &= \mathcal{C}_i(\imath)Y^{-1}(\imath)
\end{aligned} \tag{6.19}$$

where

$$\begin{aligned}
\mathcal{M}_{ij}(\imath) = &-A_i^T(\imath) - X(\imath)A_i(\imath)Y(\imath) - \left[Y^{-1}(\imath) - X(\imath)\right]\hat{B}_i(\imath)C_{2_j}(\imath)Y(\imath) \\
&- \textstyle\sum_{k=1}^s \lambda_{\imath k}Y^{-1}(k)Y(\imath) - \tilde{C}_{1_i}^T(\imath)\left[\tilde{C}_{1_j}(\imath)Y(\imath) + \tilde{D}_{12}(\imath)\hat{C}_j(\imath)Y(\imath)\right] \\
&-\gamma^{-2}\left\{X(\imath)\tilde{B}_{1_i}(\imath) + \left[Y^{-1}(\imath) - X(\imath)\right]\hat{B}_i(\imath)\tilde{D}_{21_i}(\imath)\right\}\tilde{B}_{1_j}^T(\imath).
\end{aligned} \tag{6.20}$$

Proof: Suppose there exist $X(\imath)$ and $Y(\imath)$ such that the inequalities (6.10) and (6.11)-(6.12) hold. The inequality (6.10) implies that the matrix $P(\imath)$ defined in (6.7) is a positive definite matrix. Using the partition (6.8), the filter (6.9) and multiplying (6.7) to the left by $\begin{bmatrix} Y(\imath) & I \\ Y(\imath) & 0 \end{bmatrix}$ and to the right by $\begin{bmatrix} Y(\imath) & Y(\imath) \\ I & 0 \end{bmatrix}$, we have

$$\begin{bmatrix} \Phi_{11_{ij}}(\imath) & 0 \\ 0 & \Phi_{22_{ij}}(\imath) \end{bmatrix} \tag{6.21}$$

where

$$\begin{aligned}\Phi_{11_{ij}}(\imath) &= A_i(\imath)Y(\imath) + Y(\imath)A_i^T(\imath) + \lambda_{\imath\imath}Y(\imath) + \gamma^{-2}\tilde{B}_{1_i}(\imath)\tilde{B}_{1_j}^T(\imath) \\ &\quad +[Y(\imath)\tilde{C}_{1_i}^T(\imath) + C_i^T(\imath)\tilde{D}_{12}^T(\imath)][Y(\imath)\tilde{C}_{1_i}^T(\imath) + C_i^T(\imath)\tilde{D}_{12}^T(\imath)]^T \\ &\quad +\mathcal{J}(\imath)\mathcal{Y}^{-1}(\imath)\mathcal{J}^T(\imath) \qquad (6.22)\end{aligned}$$

$$\begin{aligned}\Phi_{22_{ij}}(\imath) &= A_i^T(\imath)X(\imath) + X(\imath)A_i(\imath) + \mathcal{B}_i(\imath)C_{2_j}(\imath) + C_{2_i}^T(\imath)\mathcal{B}_j^T(\imath) \\ &\quad +\gamma^{-2}\Big[X(\imath)\tilde{B}_{1_i}(\imath) + \mathcal{B}_i(\imath)\tilde{D}_{21_j}(\imath)\Big]\Big[\tilde{B}_{1_i}^T(\imath)X(\imath) + \tilde{D}_{21_i}^T(\imath)\mathcal{B}_i^T(\imath)\Big] \\ &\quad +\tilde{C}_{1_i}^T(\imath)\tilde{C}_{1_j}(\imath) + \sum_{k=1}^{s}\lambda_{\imath k}X(k). \qquad (6.23)\end{aligned}$$

Note that $\Phi_{11_{ij}}$ and $\Phi_{22_{ij}}$ are the Schur complements of $\Psi_{11_{ij}}$ and $\Psi_{22_{ij}}$. Using (6.13)-(6.16), we have (6.21) less than zero. Hence, by Theorem 9, we learn that the inequality (6.3) holds. ■

6.1.2 Case II–$\nu(t)$ is unavailable for feedback

Now, the premise variable of the fuzzy model $\nu(t)$ is unavailable for feedback which implies μ_i is unavailable for feedback. Hence, we cannot select our filter which depends on μ_i. Thus, we select our filter as follows [91]:

$$\begin{aligned}\dot{\hat{x}}(t) &= \textstyle\sum_{i=1}^{r}\sum_{j=1}^{r}\hat{\mu}_i\hat{\mu}_j\Big[\hat{A}_{ij}(\imath)\hat{x}(t) + \hat{B}_i(\imath)y(t)\Big] \\ \hat{z}(t) &= \textstyle\sum_{i=1}^{r}\hat{\mu}_i\hat{C}_i(\imath)\hat{x}(t)\end{aligned} \qquad (6.24)$$

where $\hat{\mu}_i$ depends on the premise variable of the filter which is different from μ_i.

By applying the same technique used in Subsection 3.3.2, we have the following theorem.

Theorem 10. *Consider the system (6.1). Given a prescribed $\mathcal{H}_\infty$ performance $\gamma > 0$ and any positive constants $\delta(\imath)$, for $\imath = 1, 2, \cdots, s$, if there exist matrices $X(\imath) = X^T(\imath)$, $Y(\imath) = Y^T(\imath)$, $\mathcal{B}_i(\imath)$ and $\mathcal{C}_i(\imath)$, $i = 1, 2, \cdots, r$, satisfying the following linear matrix inequalities:*

$$\begin{bmatrix} X(\imath) & I \\ I & Y(\imath)\end{bmatrix} > 0 \qquad (6.25)$$

$$X(\imath) > 0 \qquad (6.26)$$

$$Y(\imath) > 0 \qquad (6.27)$$

$$\Psi_{11_{ii}}(\imath) < 0, \quad i = 1, 2, \cdots, r \qquad (6.28)$$

$$\Psi_{22_{ii}}(\imath) < 0, \quad i = 1, 2, \cdots, r \qquad (6.29)$$

$$\Psi_{11_{ij}}(\imath) + \Psi_{11_{ji}}(\imath) < 0, \quad i < j \le r \qquad (6.30)$$

$$\Psi_{22_{ij}}(\imath) + \Psi_{22_{ji}}(\imath) < 0, \quad i < j \le r \qquad (6.31)$$

where

$$\Psi_{11_{ij}}(\imath) = \begin{pmatrix} \begin{pmatrix} A_i(\imath)Y(\imath) + Y(\imath)A_i^T(\imath) \\ +\lambda_{\imath\imath}Y(\imath) + \gamma^{-2}\tilde{\bar{B}}_{1_i}(\imath)\tilde{\bar{B}}_{1_j}^T(\imath) \end{pmatrix} & (*)^T & (*)^T \\ \tilde{\bar{C}}_{1_i}(\imath)Y(\imath) + \tilde{\bar{D}}_{12}(\imath)\mathcal{C}_j(\imath) & -I & (*)^T \\ \mathcal{J}^T(\imath) & 0 & -\mathcal{Y}(\imath) \end{pmatrix} \tag{6.32}$$

$$\Psi_{22_{ij}}(\imath) = \begin{pmatrix} \begin{pmatrix} A_i^T(\imath)X(\imath) + X(\imath)A_i(\imath) \\ +\mathcal{B}_i(\imath)C_{2_j}(\imath) + C_{2_i}^T(\imath)\mathcal{B}_j^T(\imath) \\ +\tilde{\bar{C}}_{1_i}^T(\imath)\tilde{\bar{C}}_{1_j}(\imath) + \sum_{k=1}^{s}\lambda_{\imath k}X(k) \end{pmatrix} & (*)^T \\ \tilde{\bar{B}}_{1_i}^T(\imath)X(\imath) + \tilde{\bar{D}}_{21_i}^T(\imath)\mathcal{B}_j^T(\imath) & -\gamma^2 I \end{pmatrix} \tag{6.33}$$

with

$$\begin{aligned}
\mathcal{J}(\imath) &= \left[\sqrt{\lambda_{1\imath}}Y(\imath) \ \cdots \ \sqrt{\lambda_{(\imath-1)\imath}}Y(\imath) \ \sqrt{\lambda_{(\imath+1)\imath}}Y(\imath) \ \cdots \ \sqrt{\lambda_{s\imath}}Y(\imath)\right] \\
\mathcal{Y}(\imath) &= diag\left\{Y(1), \ \cdots, \ Y(\imath-1), \ Y(\imath+1), \ \cdots \ ,Y(s)\right\} \\
\tilde{\bar{B}}_{1_i}(\imath) &= [\delta(\imath)I \ I \ 0 \ B_{1_i}(\imath) \ 0] \\
\tilde{\bar{C}}_{1_i}(\imath) &= \left[\tfrac{\gamma\bar{\rho}(\imath)}{\delta(\imath)}\bar{H}_{1_i}^T(\imath) \ \tfrac{\gamma\bar{\rho}(\imath)}{\delta(\imath)}\bar{H}_{5_i}^T(\imath) \ \sqrt{2}\bar{\aleph}(\imath)\bar{\rho}(\imath)\bar{H}_{4_i}^T(\imath) \ \sqrt{2}\bar{\aleph}(\imath)C_{1_i}^T(\imath)\right]^T \\
\tilde{\bar{D}}_{12}(\imath) &= \left[0 \ 0 \ 0 \ -\sqrt{2}\bar{\aleph}(\imath)I\right]^T \\
\tilde{\bar{D}}_{21_i}(\imath) &= [0 \ 0 \ \delta(\imath)I \ D_{21_i}(\imath) \ I] \\
\bar{\aleph}(\imath) &= \left(1 + \bar{\rho}^2(\imath)\sum_{i=1}^{r}\sum_{j=1}^{r}\left[\|\bar{H}_{2_i}^T(\imath)\bar{H}_{2_j}(\imath)\| + \|\bar{H}_{7_i}^T(\imath)\bar{H}_{7_j}(\imath)\|\right]\right)^{\frac{1}{2}},
\end{aligned}$$

then the prescribed $\mathcal{H}_\infty$ performance $\gamma > 0$ is guaranteed. Furthermore, a suitable filter is of the form (6.24) with

$$\begin{aligned}
\hat{A}_{ij}(\imath) &= \left[Y^{-1}(\imath) - X(\imath)\right]^{-1}\mathcal{M}_{ij}(\imath)Y^{-1}(\imath) \\
\hat{B}_i(\imath) &= \left[Y^{-1}(\imath) - X(\imath)\right]^{-1}\mathcal{B}_i(\imath) \\
\hat{C}_i(\imath) &= \mathcal{C}_i(\imath)Y^{-1}(\imath)
\end{aligned} \tag{6.34}$$

where

$$\begin{aligned}
\mathcal{M}_{ij}(\imath) = &-A_i^T(\imath) - X(\imath)A_i(\imath)Y(\imath) - \left[Y^{-1}(\imath) - X(\imath)\right]\hat{B}_i(\imath)C_{2_j}(\imath)Y(\imath) \\
&- \textstyle\sum_{k=1}^{s}\lambda_{\imath k}Y^{-1}(k)Y(\imath) - \tilde{\bar{C}}_{1_i}^T(\imath)\left[\tilde{\bar{C}}_{1_j}(\imath)Y(\imath) + \tilde{\bar{D}}_{12}(\imath)\hat{C}_j(\imath)Y(\imath)\right] \\
&-\gamma^{-2}\left\{X(\imath)\tilde{\bar{B}}_{1_i}(\imath) + \left[Y^{-1}(\imath) - X(\imath)\right]\hat{B}_i(\imath)\tilde{\bar{D}}_{21_i}(\imath)\right\}\tilde{\bar{B}}_{1_j}^T(\imath).
\end{aligned} \tag{6.35}$$

Proof: It can be shown by employing the same technique used in the proof for Theorem 8. ■

6.2 Example

Consider the tunnel diode circuit shown in Figure 4.4 where the tunnel diode is characterized by

$$i_D(t) = 0.002v_D(t) + \alpha v_D^3(t)$$

where α is the characteristic parameter. The circuit is governed by the following state equations:

$$\begin{aligned} C\dot{x}_1(t) &= -0.002x_1(t) - \alpha x_1^3(t) + x_2(t) \\ L\dot{x}_2(t) &= -x_1(t) - Rx_2(t) + 0.1w_2(t) \\ y(t) &= Jx(t) + 0.1w_1(t) \\ z(t) &= \begin{bmatrix} x_1(t) \\ x_2(t) \end{bmatrix} \end{aligned} \tag{6.36}$$

where $w(t)$ is the disturbance noise input, $y(t)$ is the measurement output, $z(t)$ is the state to be estimated and J is the sensor matrix. Note that the variables $x_1(t)$ and $x_2(t)$ are the deviation variables (variables deviate from the desired trajectories). The parameters in the circuit are given as follows: $C = 20\ mF$, $L = 1000\ mH$ and $R = 10\ \Omega$. Suppose that this system is aggregated into 3 modes as shown in Table 6.1:

Table 6.1. System Terminology.

Mode $\imath$	$\alpha(\imath) \pm \Delta\alpha(\imath)$
1	0.01 ±10%
2	0.02 ±10%
3	0.03 ±10%

with the nominal transition probability matrix that relates the three operation modes

$$P_{\imath k} = \begin{bmatrix} 0.67\ 0.17\ 0.16 \\ 0.30\ 0.47\ 0.23 \\ 0.26\ 0.10\ 0.64 \end{bmatrix}.$$

With these parameters, (6.36) can be rewritten as

$$\begin{aligned} \dot{x}_1(t) &= -0.1x_1(t) - \left(\frac{[\alpha(\imath)+\Delta\alpha(\imath)]}{C}x_1^2(t)\right) \cdot x_1(t) + 50x_2(t) \\ \dot{x}_2(t) &= -x_1(t) - 10x_2(t) + 0.1w_2(t) \\ y(t) &= Jx(t) + 0.1w_1(t) \\ z(t) &= \begin{bmatrix} x_1(t) \\ x_2(t) \end{bmatrix}. \end{aligned} \tag{6.37}$$

For the sake of simplicity, we will use as few rules as possible. Assuming that $|x_1(t)| \leq 3$, the nonlinear network system (6.37) can be approximated by the following TS fuzzy model:

Plant Rule 1: IF $x_1(t)$ is $M_1(x_1(t))$ THEN

$$\begin{aligned}
\dot{x}(t) &= [A_1(\imath) + \Delta A_1(\imath)]x(t) + B_{1_1}(\imath)w(t), \quad x(0) = 0, \\
z(t) &= C_{1_1}(\imath)x(t), \\
y(t) &= C_{2_1}(\imath)x(t) + D_{21_1}(\imath)w(t).
\end{aligned}$$

Plant Rule 2: IF $x_1(t)$ is $M_2(x_1(t))$ THEN

$$\begin{aligned}
\dot{x}(t) &= [A_2(\imath) + \Delta A_2(\imath)]x(t) + B_{1_2}(\imath)w(t), \quad x(0) = 0, \\
z(t) &= C_{1_2}(\imath)x(t), \\
y(t) &= C_{2_2}(\imath)x(t) + D_{21_2}(\imath)w(t)
\end{aligned}$$

where

$$A_1(1) = \begin{bmatrix} -0.1 & 50 \\ -1 & -10 \end{bmatrix}, \; A_2(1) = \begin{bmatrix} -4.6 & 50 \\ -1 & -10 \end{bmatrix},$$

$$A_1(2) = \begin{bmatrix} -0.1 & 50 \\ -1 & -10 \end{bmatrix}, \; A_2(2) = \begin{bmatrix} -9.1 & 50 \\ -1 & -10 \end{bmatrix},$$

$$A_1(3) = \begin{bmatrix} -0.1 & 50 \\ -1 & -10 \end{bmatrix}, \; A_2(3) = \begin{bmatrix} -13.6 & 50 \\ -1 & -10 \end{bmatrix},$$

$$B_{1_1}(\imath) = B_{1_2}(\imath) = \begin{bmatrix} 0 & 0 \\ 0 & 0.1 \end{bmatrix}, \; C_{1_1}(\imath) = C_{1_2}(\imath) = \begin{bmatrix} 1 & 0 \\ 0 & 1 \end{bmatrix},$$

$$C_{2_1}(\imath) = C_{2_2}(\imath) = J, \; D_{21_1}(\imath) = D_{21_2}(\imath) = \begin{bmatrix} 0.1 & 0 \end{bmatrix},$$

$$\Delta A_1(\imath) = F(x(t), \imath, t)H_{1_1}(\imath) \text{ and } A_2(\imath) = F(x(t), \imath, t)H_{1_2}(\imath).$$

Now, by assuming that $\|F(x(t), \imath, t)\| \leq \rho(\imath) = 1$, we have

$$H_{1_1}(1) = \begin{bmatrix} 0 & 0 \\ 0 & 0 \end{bmatrix}, \; H_{1_2}(1) = \begin{bmatrix} -0.45 & 0 \\ 0 & 0 \end{bmatrix},$$

$$H_{1_1}(2) = \begin{bmatrix} 0 & 0 \\ 0 & 0 \end{bmatrix}, \; H_{1_2}(2) = \begin{bmatrix} -0.9 & 0 \\ 0 & 0 \end{bmatrix},$$

$$H_{1_1}(3) = \begin{bmatrix} 0 & 0 \\ 0 & 0 \end{bmatrix} \text{ and } H_{1_2}(3) = \begin{bmatrix} -1.35 & 0 \\ 0 & 0 \end{bmatrix}.$$

Note that the plot of the membership function Rules 1 and 2 is the same as in Figure 4.5.

Case I-$\nu(t)$ is available for feedback

In this case, $x_1(t) = \nu(t)$ is assumed to be available for feedback; for instance, $J = [1 \;\; 0]$. This implies that μ_i is available for feedback. Using the LMI optimization algorithm and Theorem 6.1 with $\gamma = 1$ and $\delta(1) = \delta(2) = \delta(3) = 1$, we obtain

$$X(1) = \begin{bmatrix} 1.3527 & 4.1536 \\ 4.1536 & 23.7154 \end{bmatrix}, \qquad Y(1) = \begin{bmatrix} 15.9976 & -0.2409 \\ -0.2409 & 0.5000 \end{bmatrix},$$

$$\hat{A}_{11}(1) = \begin{bmatrix} -50.5324 & -1.7600 \\ -9.7924 & -0.5462 \end{bmatrix}, \qquad \hat{A}_{12}(1) = \begin{bmatrix} -50.5324 & -1.7600 \\ -9.7924 & -0.5462 \end{bmatrix},$$

$$\hat{A}_{21}(1) = \begin{bmatrix} -53.3639 & -1.8542 \\ -19.4469 & -0.3911 \end{bmatrix}, \qquad \hat{A}_{22}(1) = \begin{bmatrix} -53.3639 & -1.8542 \\ -19.4469 & -0.3911 \end{bmatrix},$$

$$\hat{B}_1(1) = \begin{bmatrix} 0.2743 \\ -0.9846 \end{bmatrix}, \qquad \hat{B}_2(1) = \begin{bmatrix} 0.3067 \\ -1.2423 \end{bmatrix},$$

$$\hat{C}_1(1) = \begin{bmatrix} -35.3553 & -1.1213 \end{bmatrix}, \qquad \hat{C}_2(1) = \begin{bmatrix} -35.3553 & 0.1110 \end{bmatrix},$$

$$X(2) = \begin{bmatrix} 1.1422 & 3.3069 \\ 3.3069 & 19.7273 \end{bmatrix}, \qquad Y(2) = \begin{bmatrix} 8.8351 & -0.1880 \\ -0.1880 & 0.3363 \end{bmatrix},$$

$$\hat{A}_{11}(2) = \begin{bmatrix} -52.3064 & -2.3475 \\ -3.8388 & -0.5670 \end{bmatrix}, \qquad \hat{A}_{12}(2) = \begin{bmatrix} -52.3064 & -2.3475 \\ -3.8388 & -0.5670 \end{bmatrix},$$

$$\hat{A}_{21}(2) = \begin{bmatrix} -58.4742 & -2.4526 \\ -25.9706 & -0.1006 \end{bmatrix}, \qquad \hat{A}_{22}(2) = \begin{bmatrix} -58.4742 & -2.4526 \\ -25.9706 & -0.1006 \end{bmatrix},$$

$$\hat{B}_1(2) = \begin{bmatrix} 0.4488 \\ -1.6417 \end{bmatrix}, \qquad \hat{B}_2(2) = \begin{bmatrix} 0.0851 \\ -0.5918 \end{bmatrix},$$

$$\hat{C}_1(2) = \begin{bmatrix} -35.3553 & -0.1998 \end{bmatrix}, \qquad \hat{C}_2(2) = \begin{bmatrix} -35.3553 & -0.2554 \end{bmatrix},$$

$$X(3) = \begin{bmatrix} 0.9146 & 2.5472 \\ 2.5472 & 16.0807 \end{bmatrix}, \qquad Y(3) = \begin{bmatrix} 5.8540 & -0.1805 \\ -0.1805 & 0.2579 \end{bmatrix},$$

$$\hat{A}_{11}(3) = \begin{bmatrix} -53.3336 & -2.8124 \\ -0.7319 & -0.7547 \end{bmatrix}, \qquad \hat{A}_{12}(3) = \begin{bmatrix} -53.3336 & -2.8124 \\ -0.7319 & -0.7547 \end{bmatrix},$$

$$\hat{A}_{21}(3) = \begin{bmatrix} -63.4126 & -3.1736 \\ -22.7881 & -0.0209 \end{bmatrix}, \qquad \hat{A}_{22}(3) = \begin{bmatrix} -63.4126 & -3.1736 \\ -22.7881 & -0.0209 \end{bmatrix},$$

$$\hat{B}_1(3) = \begin{bmatrix} 0.7630 \\ -2.9262 \end{bmatrix}, \qquad \hat{B}_2(3) = \begin{bmatrix} 0.0795 \\ -0.7686 \end{bmatrix},$$

$$\hat{C}_1(3) = \begin{bmatrix} -35.3553 & -1.6653 \end{bmatrix}, \qquad \hat{C}_2(3) = \begin{bmatrix} -35.3553 & 0.2665 \end{bmatrix}.$$

The resulting fuzzy filter is

$$\begin{aligned} \dot{\hat{x}}(t) &= \textstyle\sum_{i=1}^{2} \sum_{j=1}^{2} \mu_i \mu_j \hat{A}_{ij}(\imath)\hat{x}(t) + \sum_{i=1}^{2} \mu_i \hat{B}_i(\imath) y(t) \\ \hat{z}(t) &= \textstyle\sum_{i=1}^{2} \mu_i \hat{C}_i(\imath)\hat{x}(t) \end{aligned} \tag{6.38}$$

where

$$\mu_1 = M_1(x_1(t)) \text{ and } \mu_2 = M_2(x_1(t)).$$

Case II-$\nu(t)$ is unavailable for feedback

In this case, $x_1(t) = \nu(t)$ is assumed to be unavailable for feedback; for instance, $J = [0 \;\; 1]$. This implies that μ_i is unavailable for feedback. Using the LMI optimization algorithm and Theorem 6.2 with $\gamma = 1$ and $\delta(1) = \delta(2) = \delta(3) = 1$, we obtain

$$X(1) = \begin{bmatrix} 1.3721 & 4.2243 \\ 4.2243 & 24.1080 \end{bmatrix}, \qquad Y(1) = \begin{bmatrix} 14.7533 & -0.2063 \\ -0.2063 & 0.4399 \end{bmatrix},$$

$$\hat{A}_{11}(1) = \begin{bmatrix} -50.7139 & -1.7308 \\ -22.5449 & -0.0146 \end{bmatrix}, \qquad \hat{A}_{12}(1) = \begin{bmatrix} -50.7139 & -1.7308 \\ -22.5449 & -0.0146 \end{bmatrix},$$

$$\hat{A}_{21}(1) = \begin{bmatrix} -53.6150 & -1.7741 \\ -24.4667 & -0.8441 \end{bmatrix}, \qquad \hat{A}_{22}(1) = \begin{bmatrix} -53.6150 & -1.7741 \\ -24.4667 & -0.8441 \end{bmatrix},$$

$$\hat{B}_1(1) = \begin{bmatrix} 0.1802 \\ -0.7387 \end{bmatrix}, \qquad \hat{B}_2(1) = \begin{bmatrix} 0.5358 \\ -1.8729 \end{bmatrix},$$

$$\hat{C}_1(1) = \begin{bmatrix} -35.3553 & 1.0222 \end{bmatrix}, \qquad \hat{C}_2(1) = \begin{bmatrix} -35.3553 & 0.1221 \end{bmatrix},$$

$$X(2) = \begin{bmatrix} 1.1564 & 3.3632 \\ 3.3632 & 20.0687 \end{bmatrix}, \qquad Y(2) = \begin{bmatrix} 8.1386 & -0.1553 \\ -0.1553 & 0.2925 \end{bmatrix},$$

$$\hat{A}_{11}(2) = \begin{bmatrix} -52.9363 & -2.1627 \\ -15.3598 & 0.3097 \end{bmatrix}, \qquad \hat{A}_{12}(2) = \begin{bmatrix} -52.9363 & -2.1627 \\ -15.3598 & 0.3097 \end{bmatrix},$$

$$\hat{A}_{21}(2) = \begin{bmatrix} -59.2867 & -2.2823 \\ -28.1564 & -0.9187 \end{bmatrix}, \qquad \hat{A}_{22}(2) = \begin{bmatrix} -59.2867 & -2.2823 \\ -28.1564 & -0.9187 \end{bmatrix},$$

$$\hat{B}_1(2) = \begin{bmatrix} 0.5723 \\ -1.6360 \end{bmatrix}, \qquad \hat{B}_2(2) = \begin{bmatrix} 1.0338 \\ -1.7367 \end{bmatrix},$$

$$\hat{C}_1(2) = \begin{bmatrix} -35.3553 & -1.4211 \end{bmatrix}, \qquad \hat{C}_2(2) = \begin{bmatrix} -35.3553 & 0 \end{bmatrix},$$

$$X(3) = \begin{bmatrix} 0.9254 & 2.5969 \\ 2.5969 & 16.4096 \end{bmatrix}, \qquad Y(3) = \begin{bmatrix} 5.4341 & -0.1491 \\ -0.1491 & 0.2228 \end{bmatrix},$$

$$\hat{A}_{11}(3) = \begin{bmatrix} -54.8946 & -2.9091 \\ -12.0349 & 0.4766 \end{bmatrix}, \qquad \hat{A}_{12}(3) = \begin{bmatrix} -54.8946 & -2.9091 \\ -12.0349 & 0.4766 \end{bmatrix},$$

$$\hat{A}_{21}(3) = \begin{bmatrix} -64.5265 & -2.7580 \\ -24.0698 & -1.2716 \end{bmatrix}, \qquad \hat{A}_{22}(3) = \begin{bmatrix} -64.5265 & -2.7580 \\ -24.0698 & -1.2716 \end{bmatrix},$$

$$\hat{B}_1(3) = \begin{bmatrix} 1.0373 \\ -1.1620 \end{bmatrix}, \qquad \hat{B}_2(3) = \begin{bmatrix} 1.1281 \\ -1.5550 \end{bmatrix},$$

$$\hat{C}_1(3) = \begin{bmatrix} -35.3553 & 1.4877 \end{bmatrix}, \qquad \hat{C}_2(3) = \begin{bmatrix} -35.3553 & -0.3775 \end{bmatrix}.$$

The resulting fuzzy filter is

$$\begin{aligned} \dot{\hat{x}}(t) &= \textstyle\sum_{i=1}^{2}\sum_{j=1}^{2}\hat{\mu}_i\hat{\mu}_j\hat{A}_{ij}(\imath)\hat{x}(t) + \sum_{i=1}^{2}\hat{\mu}_i\hat{B}_i(\imath)y(t) \\ \hat{z}(t) &= \textstyle\sum_{i=1}^{2}\hat{\mu}_i\hat{C}_i(\imath)\hat{x}(t) \end{aligned} \tag{6.39}$$

where

$$\hat{\mu}_1 = M_1(\hat{x}_1(t)) \text{ and } \hat{\mu}_2 = M_2(\hat{x}_1(t)).$$

Remark 4. Figures 6.1(a)-6.1(b), respectively, show the responses of $x_1(t)$ *and* $x_2(t)$ *in Cases I and II. Figure 6.2 shows the result of the changing between modes during the simulation with the initial mode 2. The disturbance input signal,* $w(t)$*, which was used during the simulation is given in Figure 6.3. The simulation results for the ratio of the filter error energy to the disturbance input noise energy obtained by using the* $\mathcal{H}_\infty$ *fuzzy filter are depicted in Figure 6.4. After* 15 *seconds, the ratio of the filter error energy to the disturbance input noise energy tends to a constant value which is about* 0.33 *in Case I and* 0.38 *in Case II. Thus, in Case I where* $\gamma = \sqrt{0.33} = 0.574$ *and in Case II where* $\gamma = \sqrt{0.38} = 0.616$*, both are less than the prescribed value* 1. □

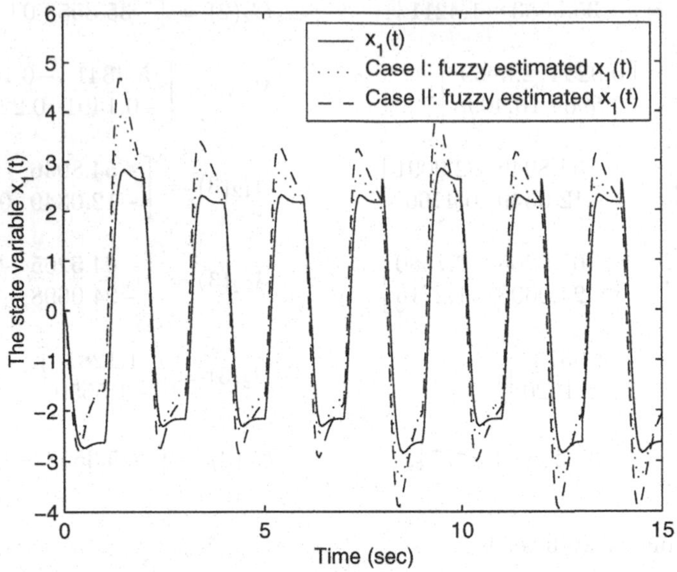

(a) The histories of $x_1(t)$

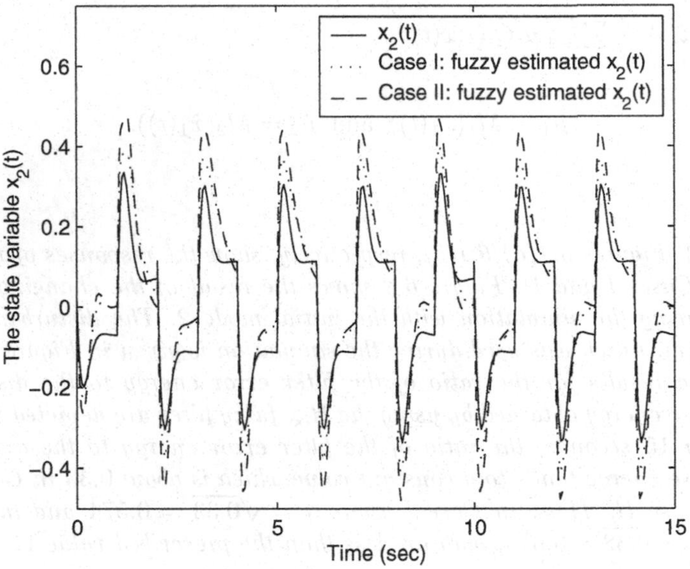

(b) The histories of $x_2(t)$

Fig. 6.1. The histories of $x_1(t)$ and $x_2(t)$ in Cases I and II.

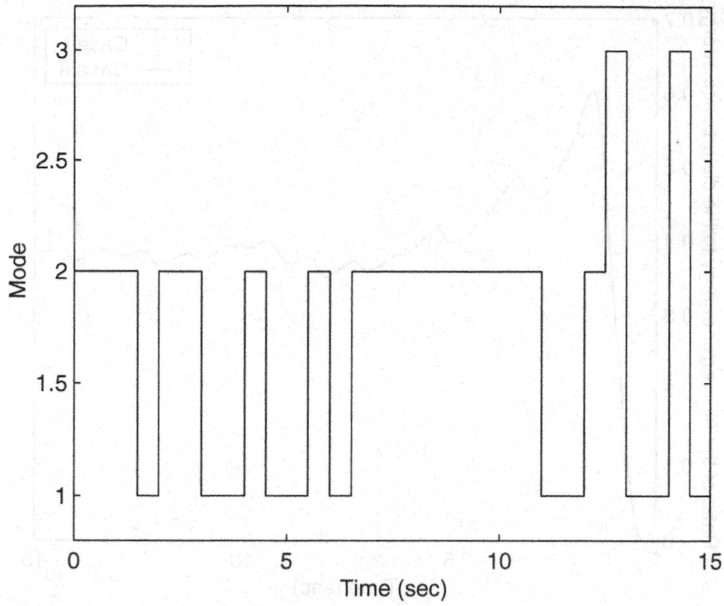

Fig. 6.2. The result of the changing between modes during the simulation with the initial mode 2.

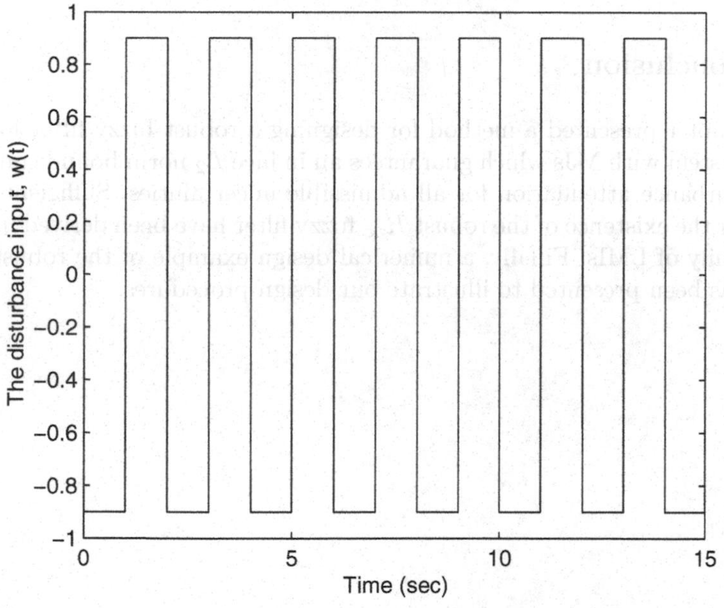

Fig. 6.3. The disturbance input noise, $w(t)$.

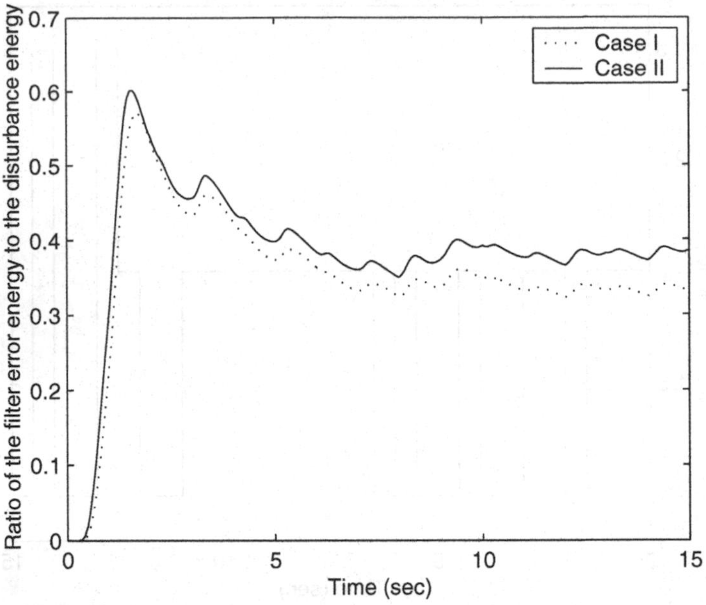

Fig. 6.4. The ratio of the filter error energy to the disturbance noise energy: $\left(\frac{\int_0^{T_f}(z(t)-\hat{z}(t))^T(z(t)-\hat{z}(t))dt}{\int_0^{T_f}w^T(t)w(t)dt}\right)$.

6.3 Conclusion

This chapter presented a method for designing a robust fuzzy filter for a TS fuzzy system with MJs which guarantees an induce $\mathcal{L}_2$ norm bound constraint on disturbance attenuation for all admissible uncertainties. Sufficient conditions for the existence of the robust $\mathcal{H}_\infty$ fuzzy filter have been derived in terms of a family of LMIs. Finally, a numerical design example of the robust fuzzy filter has been presented to illustrate our design procedures.

Part II

UNCERTAIN FUZZY SINGULARLY PERTURBED SYSTEMS

7

Uncertain Fuzzy Singularly Perturbed Systems

7.1 Background and Motivation

Singularly perturbed systems (SPSs), sometimes called multiple time-scale dynamic systems, normally occur due to the presence of small "parasitic" parameters such as small time constants and masses. Examples of SPSs can be found in every discipline. In power system models, a small "parasitic" parameter can represent machine reactance or transients in voltage regulators. In industrial control systems, it may represent time constants of drives and actuators. In biochemical models, a small "parasitic" parameter can indicate a small quantity of an enzyme. In flexible booster models, a small "parasitic" parameter is due to bending modes and in nuclear reactor models, it is due to fast neutrons. The presence of these "parasitic" parameters can make the dimensionality of a dynamics system prohibitively high.

For the past three decades, SPSs have been intensively studied by many researchers; e.g., [93, 94, 95, 96, 97, 98, 99, 100, 101, 102, 103, 104, 105, 106, 107, 108, 109, 110, 111, 112, 113, 114, 95, 100, 115, 116, 117, 118, 119, 120, 121]. The main purpose of the singular perturbation approach to analysis and design is the alleviation of high dimensionality and ill-conditioning which is caused by the interaction of slow and fast dynamic modes. The separation of states into slow and fast subsystems is a nontrivial modelling task demanding insight and ingenuity on the part of the analyst. In state space, such systems are commonly modelled using the mathematical framework of singular perturbations, with a small parameter, say ε, determining the degree of separation between the "slow" and "fast" modes of the system. However, it is necessary to note that it is possible to solve SPSs without separating into slow and fast mode subsystems if the "parasitic" parameters are large enough. In the case of having very small "parasitic" parameters which normally occur in the description of various physical phenomena, a popular approach adopted to handle these systems is based on the so-called reduction technique [104]. According to this technique, the fast variables are replaced by their steady states obtained with "frozen" slow variables and controls, and the slow dynamics are approximated by the corresponding reduced order system. This time-scale is asymptotic, that is, exact in the limit, as the ratio of the speeds of the slow versus the fast dynamics tends to zero.

Some of the first works in the area of SPSs were investigated by Tikhonov, [122], in 1948. This work related to the solution of differential equations with a small parameter multiplying the derivative. Later, modern control theories were introduced in the 1960's by many researchers such as Bellman, Kalman and others, which could be easily applied to the singularly perturbed problems; e.g., [123, 124, 125]. As a result, the last 43 years has witnessed a rapid growth in further development of these control methods for SPSs. Literature surveys on singular perturbation, [103, 126], list over 350 post-1980's references on this topic.

In 1969, the singularly perturbed regulator problem was originally posed by Sannuti and Kokotovic [102]. Their works were based on the condition of assumed stability of certain partitions of the plant matrix and no fast output terms. Later, the papers of Chow and Kokotovic [127] and Suzuki and Miura [128] addressed control for the regulator and state-feedback stabilization problem. The theory of composite feedback control was developed further by Khalil, [101], in 1987. In recent decades, the research on SPSs in the $\mathcal{H}_\infty$ sense has been highly recognized in the control area due to the great practical importance. For examples, Pan and Basar, [106, 107], investigated the $\mathcal{H}_\infty$-optimal control of SPSs under both perfect and imperfect state measurement. Shi and Dragan, [110, 111], have considered the asymptotic $\mathcal{H}_\infty$ control design of SPSs with parametric uncertainties. In [109], the authors have investigated the decomposition solution of $\mathcal{H}_\infty$ filter gain for SPSs, while the reduced-order $\mathcal{H}_\infty$ optimal filtering for system with slow and fast modes has been considered in [93]. Recently, Dragan, [129], has developed an $\mathcal{H}_\infty$ controller for LSPS–MJs via the slow-fast decomposition approach.

Although many researchers have studied the $\mathcal{H}_\infty$ control and filter design of LSPSs and LSPS–MJs for many years, the $\mathcal{H}_\infty$ control and filter design of nonlinear singularly perturbed systems (NSPSs) and nonlinear singularly perturbed systems with Markovian jumps (NSPS–MJs) remains as open research areas. This is generally due to the fact that NSPSs and NSPS-MJs cannot be decomposed into slow and fast subsystems. Furthermore, if we employ the existing fuzzy results in [34, 35, 27, 38, 39, 40, 41, 42, 43, 44, 48, 49, 50, 51, 52, 53] on the NSPS and NSPS–MJs, they end up with a family of ill-conditioned linear matrix inequalities (LMIs) resulting from the interaction of slow and fast dynamic modes. In general, ill-conditioned LMIs are very difficult to solve. Recently, there have been some attempts in $\mathcal{H}_\infty$ control for a class of NSPSs, however, nonlinearity only on the slow variables has been examined; e.g., [98, 130, 131, 132, 133, 134, 135]. A local state-feedback $\mathcal{H}_\infty$ control problem for affine NSPSs has been also addressed in [98]. In [136], a global state-feedback $\mathcal{H}_\infty$ control problem for a class of NSPSs described by the TS fuzzy model has been studied. So far, to the best of our knowledge, an LMI approach to the $\mathcal{H}_\infty$ control and filter problem of the parametric uncertainties issue in NSPS and NSPS–MJ based on the TS fuzzy model has not yet been considered in the literature.

This motivates us to consider the topics in UNSPSs and UNSPS–MJs since in general, UNSPSs and UNSPS–MJs cannot be decomposed into slow and fast subsystems. The commonly used method to solve the design problem for these classes of systems is basically based on a linearized technique and a slow-fast decomposition approach which may result in a system far from its desired performance. Thus, it is necessary to have an approach that can help us to overcome the design difficulties of the controller and filter for a class of UNSPSs and UNSPS–MJs.

Therefore, in order to bridge the gap, this book will present a new novel methodology on designing a robust $\mathcal{H}_\infty$ fuzzy controller and a robust $\mathcal{H}_\infty$ fuzzy filter for a class of UNSPSs and UNSPS–MJs which are described by a TS fuzzy system with parametric uncertainties and with/without MJs. The proposed design approach in this book for a class of UFSPSs and UFSPS–MJs does not involve the separation of states into slow and fast ones and it can be applied not only to standard, but also to nonstandard nonlinear singularly perturbed systems. The outline of presentation for Part II is given in next section.

7.2 Outline of Part II

In Part II, the synthesis design procedure of a robust $\mathcal{H}_\infty$ fuzzy controller and a robust $\mathcal{H}_\infty$ fuzzy filter for a class of UNSPSs and UNSPS–MJs which is described by a TS fuzzy system with parametric uncertainties and with/without MJs is presented. The outline of Part II is presented as follows. Chapter 7 provides some background and motivation on UFSPSs. Chapters 8 and 9 present the synthesis design procedure of a robust $\mathcal{H}_\infty$ fuzzy controller and a robust $\mathcal{H}_\infty$ fuzzy filter for the class of UFSPSs. Then, Chapters 10 and 11 respectively present the synthesis design procedure of a robust $\mathcal{H}_\infty$ fuzzy controller and a robust $\mathcal{H}_\infty$ fuzzy filter for the class of UFSPS–MJs. Finally, to illustrative the effectiveness of the design procedures, a numerical example is also given at the end of each chapter.

8

Robust $\mathcal{H}_\infty$ Fuzzy Control Design for Uncertain Fuzzy Singularly Perturbed Systems

In this chapter, we present a new designing technique for a robust fuzzy state and output feedback controller for a TS singularly perturbed fuzzy system with parametric uncertainties. A technique for designing a robust fuzzy controller such that the $\mathcal{L}_2$-gain of the mapping from the exogenous input noise to the regulated output is less than the prescribed value has been developed.

8.1 System Description

In this chapter, we generalize the TS singularly perturbed fuzzy system to represent a TS singularly perturbed fuzzy system with parametric uncertainties. As in [136], we examine a TS singularly perturbed fuzzy system with parametric uncertainties as follows:

$$\begin{aligned} E_\varepsilon \dot{x}(t) &= \textstyle\sum_{i=1}^r \mu_i(\nu(t))\Big[[A_i + \Delta A_i]x(t) + [B_{1_i} + \Delta B_{1_i}]w(t) \\ &\quad +[B_{2_i} + \Delta B_{2_i}]u(t)\Big], \quad x(0) = 0 \\ z(t) &= \textstyle\sum_{i=1}^r \mu_i(\nu(t))\Big[[C_{1_i} + \Delta C_{1_i}]x(t) + [D_{12_i} + \Delta D_{12_i}]u(t)\Big] \\ y(t) &= \textstyle\sum_{i=1}^r \mu_i(\nu(t))\Big[[C_{2_i} + \Delta C_{2_i}]x(t) + [D_{21_i} + \Delta D_{21_i}]w(t)\Big] \end{aligned} \tag{8.1}$$

where $E_\varepsilon = \begin{bmatrix} I & 0 \\ 0 & \varepsilon I \end{bmatrix}$, $\varepsilon > 0$ is the singular perturbation parameter, $\nu(t) = [\nu_1(t) \;\; \cdots \;\; \nu_\vartheta(t)]$ is the premise variable vector that may depend on states in many cases, $\mu_i(\nu(t))$ denotes the normalized time-varying fuzzy weighting functions for each rule (i.e., $\mu_i(\nu(t)) \geq 0$ and $\sum_{i=1}^r \mu_i(\nu(t)) = 1$), ϑ is the number of fuzzy sets, $x(t) \in \Re^n$ is the state vector, $u(t) \in \Re^m$ is the input, $w(t) \in \Re^p$ is the disturbance which belongs to $\mathcal{L}_2[0, \infty)$, $y(t) \in \Re^\ell$ is the measurement and $z(t) \in \Re^s$ is the controlled output, the matrices $A_i, B_{1_i}, B_{2_i}, C_{1_i}, C_{2_i}, D_{12_i}$ and D_{21_i} are of appropriate dimensions, and the matrices ΔA_i, ΔB_{1_i}, ΔB_{2_i}, ΔC_{1_i}, ΔC_{2_i}, ΔD_{12_i} and ΔD_{21_i} represent the uncertainties in the system and satisfy Assumption 3.1.

8.2 Robust $\mathcal{H}_\infty$ State-Feedback Control Design

The aim of this section is to design a robust $\mathcal{H}_\infty$ fuzzy state-feedback controller of the form

$$u(t) = \sum_{j=1}^{r} \mu_j K_j x(t) \tag{8.2}$$

where K_j is the controller gain, such that the inequality (3.3) holds. The state space form of the fuzzy system model (8.1) with the controller (8.2) is given by

$$\begin{aligned} E_\varepsilon \dot{x}(t) = \textstyle\sum_{i=1}^{r} \sum_{j=1}^{r} \mu_i \mu_j \Big[[(A_i + B_{2_i} K_j) + (\Delta A_i + \Delta B_{2_i} K_j)] x(t) \\ + [B_{1_i} + \Delta B_{1_i}] w(t) \Big], \qquad x(0) = 0. \end{aligned} \tag{8.3}$$

Sufficient conditions for the existence of a robust $\mathcal{H}_\infty$ fuzzy state-feedback controller are provided in the following lemma. The Lyapunov approach is used to derived these sufficient conditions.

Lemma 5. *Consider the system (8.1). Given a prescribed $\mathcal{H}_\infty$ performance $\gamma > 0$ and a positive constant δ, if there exist a matrix $P_\varepsilon = P_\varepsilon^T$ and matrices Y_j, $j = 1, 2, \cdots, r$, satisfying the following ε-dependent linear matrix inequalities:*

$$P_\varepsilon > 0 \tag{8.4}$$

$$\Psi_{ii}(\varepsilon) < 0, \quad i = 1, 2, \cdots, r \tag{8.5}$$

$$\Psi_{ij}(\varepsilon) + \Psi_{ji}(\varepsilon) < 0, \quad i < j \le r \tag{8.6}$$

where

$$\Psi_{ij}(\varepsilon) = \begin{pmatrix} A_i E_\varepsilon^{-1} P_\varepsilon + E_\varepsilon^{-1} P_\varepsilon A_i^T + B_{2_i} Y_j + Y_j^T B_{2_i}^T & (*)^T & (*)^T \\ \tilde{B}_{1_i}^T & -\gamma I & (*)^T \\ \tilde{C}_{1_i} E_\varepsilon^{-1} P_\varepsilon + \tilde{D}_{12_i} Y_j & 0 & -\gamma I \end{pmatrix} \tag{8.7}$$

with

$$\tilde{B}_{1_i} = \begin{bmatrix} \delta I \ I \ \delta I \ B_{1_i} \end{bmatrix}, \ \tilde{C}_{1_i} = \begin{bmatrix} \frac{\gamma\rho}{\delta} H_{1_i}^T \ 0 \ \sqrt{2}\lambda\rho H_{4_i}^T \ \sqrt{2}\lambda C_{1_i}^T \end{bmatrix}^T,$$

$$\tilde{D}_{12_i} = \begin{bmatrix} 0 \ \frac{\gamma\rho}{\delta} H_{3_i}^T \ \sqrt{2}\lambda\rho H_{6_i}^T \ \sqrt{2}\lambda D_{12_i}^T \end{bmatrix}^T, \lambda = \left(1 + \rho^2 \sum_{i=1}^{r} \sum_{j=1}^{r} \left[\| H_{2_i}^T H_{2_j} \| \right] \right)^{\frac{1}{2}},$$

then the inequality (3.3) holds. Furthermore, a suitable choice of the fuzzy controller is

$$u(t) = \sum_{j=1}^{r} \mu_j K_j(\varepsilon)x(t) \tag{8.8}$$

where

$$K_j(\varepsilon) = Y_j P_\varepsilon^{-1} E_\varepsilon. \tag{8.9}$$

Proof: The proof can be carried out by the same technique used in Theorem 1. ■

Remark 5. The LMIs given in Lemma 5 become ill-conditioned when ε is sufficiently small, which is always the case for the SPS. In general, these ill-conditioned LMIs are very difficult to solve. Thus, to alleviate these ill-conditioned LMIs, we have the following theorem which does not depend on ε. □

Theorem 11. *Consider the system (8.1). Given a prescribed $\mathcal{H}_\infty$ performance $\gamma > 0$ and a positive constant δ, if there exist a matrix P and matrices Y_j, $j = 1, 2, \cdots, r$, satisfying the following ε-independent linear matrix inequalities:*

$$EP = P^T E, \quad PD = DP, \quad EP + PD > 0 \tag{8.10}$$

$$\Psi_{ii} < 0, \quad i = 1, 2, \cdots, r \tag{8.11}$$

$$\Psi_{ij} + \Psi_{ji} < 0, \quad i < j \leq r \tag{8.12}$$

where $E = \begin{pmatrix} I & 0 \\ 0 & 0 \end{pmatrix}$, $D = \begin{pmatrix} 0 & 0 \\ 0 & I \end{pmatrix}$ *and*

$$\Psi_{ij} = \begin{pmatrix} A_i P + P^T A_i^T + B_{2_i} Y_j + Y_j^T B_{2_i}^T & (*)^T & (*)^T \\ \tilde{B}_{1_i}^T & -\gamma I & (*)^T \\ \tilde{C}_{1_i} P + \tilde{D}_{12_i} Y_j & 0 & -\gamma I \end{pmatrix} \tag{8.13}$$

with

$$\tilde{B}_{1_i} = \left[\delta I \ I \ \delta I \ B_{1_i} \right], \quad \tilde{C}_{1_i} = \left[\tfrac{\gamma\rho}{\delta} H_{1_i}^T \ 0 \ \sqrt{2}\lambda\rho H_{4_i}^T \ \sqrt{2}\lambda C_{1_i}^T \right]^T,$$

$$\tilde{D}_{12_i} = \left[0 \ \tfrac{\gamma\rho}{\delta} H_{3_i}^T \ \sqrt{2}\lambda\rho H_{6_i}^T \ \sqrt{2}\lambda D_{12_i}^T \right]^T, \lambda = \left(1 + \rho^2 \sum_{i=1}^{r}\sum_{j=1}^{r} \left[\| H_{2_i}^T H_{2_j} \| \right] \right)^{\frac{1}{2}},$$

then there exists a sufficiently small $\hat{\varepsilon} > 0$ such that the inequality (3.3) holds for $\varepsilon \in (0, \hat{\varepsilon}]$. Furthermore, a suitable choice of the fuzzy controller is

$$u(t) = \sum_{j=1}^{r} \mu_j K_j x(t) \tag{8.14}$$

where

$$K_j = Y_j P^{-1}. \tag{8.15}$$

Proof: Suppose there exists a matrix P such that the inequality (8.10) holds, then P is of the following form:

$$P = \begin{pmatrix} P_1 & 0 \\ P_2^T & P_3 \end{pmatrix} \tag{8.16}$$

with $P_1 = P_1^T > 0$ and $P_3 = P_3^T > 0$. Let

$$P_\varepsilon = E_\varepsilon (P + \varepsilon \tilde{P}) \tag{8.17}$$

with

$$\tilde{P} = \begin{pmatrix} 0 & P_2 \\ 0 & 0 \end{pmatrix}. \tag{8.18}$$

Substituting (8.16) and (8.18) into (8.17), we have

$$P_\varepsilon = \begin{pmatrix} P_1 & \varepsilon P_2 \\ \varepsilon P_2^T & \varepsilon P_3 \end{pmatrix}. \tag{8.19}$$

Clearly, $P_\varepsilon = P_\varepsilon^T$, and there exists a sufficient small $\hat{\varepsilon}$ such that for $\varepsilon \in (0, \hat{\varepsilon}]$, $P_\varepsilon > 0$. Using the matrix inversion lemma, we learn that

$$P_\varepsilon^{-1} = \left[P^{-1} + \varepsilon M_\varepsilon \right] E_\varepsilon^{-1} \tag{8.20}$$

where $M_\varepsilon = -P^{-1}\tilde{P}\left(I + \varepsilon P^{-1}\tilde{P}\right)^{-1} P^{-1}$. Substituting (8.17) and (8.20) into (8.7), we obtain

$$\Psi_{ij} + \psi_{ij} \tag{8.21}$$

where the ε-independent linear matrix Ψ_{ij} is defined in (8.13) and the ε-dependent linear matrix is

$$\psi_{ij} = \varepsilon \begin{pmatrix} A_i\tilde{P} + \tilde{P}^T A_i^T + B_{2_i} Y_{\varepsilon_j} + Y_{\varepsilon j}^T B_{2_i}^T & (*)^T & (*)^T \\ 0 & 0 & (*)^T \\ \tilde{C}_{1_i}\tilde{P} + \tilde{D}_{12_i} Y_{\varepsilon_j} & 0 & 0 \end{pmatrix} \tag{8.22}$$

where $Y_{\varepsilon_j} = K_j M_\varepsilon^{-1}$. Note that the ε-dependent linear matrix tends to zero when ε approaches zero.

Employing (8.10)-(8.13) and using the fact that for any given negative definite matrix $\mathcal{W}$, there exists an $\varepsilon > 0$ such that $\mathcal{W} + \varepsilon I < 0$, one can show that there exists a sufficiently small $\hat{\varepsilon} > 0$ such that for $\varepsilon \in (0, \hat{\varepsilon}]$, (8.5) and (8.6) hold. Since (8.4)-(8.6) hold, using Lemma 8.1, the inequality (3.3) holds for $\varepsilon \in (0, \hat{\varepsilon}]$. ■

8.3 Robust $\mathcal{H}_\infty$ Output Feedback Control Design

This section aims at designing a full order dynamic $\mathcal{H}_\infty$ fuzzy output feedback controller of the form

$$\begin{aligned} E_\varepsilon \dot{\hat{x}}(t) &= \sum_{i=1}^r \sum_{j=1}^r \hat{\mu}_i \hat{\mu}_j \left[\hat{A}_{ij}(\varepsilon)\hat{x}(t) + \hat{B}_i y(t)\right] \\ u(t) &= \sum_{i=1}^r \hat{\mu}_i \hat{C}_i \hat{x}(t) \end{aligned} \tag{8.23}$$

where $\hat{x}(t) \in \Re^n$ is the controller's state vector, $\hat{A}_{ij}(\varepsilon)$, $\hat{B}_i$ and $\hat{C}_i$ are parameters of the controller which are to be determined, and $\hat{\mu}_i$ denotes the normalized time-varying fuzzy weighting functions for each rule (i.e., $\hat{\mu}_i \geq 0$ and $\sum_{i=1}^r \hat{\mu}_i = 1$), such that the inequality (3.3) holds. Clearly, in real control problems, all of the premise variables are not necessarily measurable. Thus, we consider the designing of the robust $\mathcal{H}_\infty$ output feedback control into two cases as follows. In Subsection 8.3.1, we consider the case where the premise variable of the fuzzy model μ_i is measurable, while in Subsection 8.3.2, the premise variable which is assumed to be unmeasurable is considered.

8.3.1 Case I–$\nu(t)$ is available for feedback

The premise variable of the fuzzy model $\nu(t)$ is available for feedback which implies that μ_i is available for feedback. Thus, we can select our controller that depends on μ_i as follows:

$$\begin{aligned} E_\varepsilon \dot{\hat{x}}(t) &= \sum_{i=1}^r \sum_{j=1}^r \mu_i \mu_j \left[\hat{A}_{ij}(\varepsilon)\hat{x}(t) + \hat{B}_i y(t)\right] \\ u(t) &= \sum_{i=1}^r \mu_i \hat{C}_i \hat{x}(t). \end{aligned} \tag{8.24}$$

Before presenting our next results, the following lemma is recalled.

Lemma 6. *Consider the system (8.1). Given a prescribed $\mathcal{H}_\infty$ performance γ and a positive constant δ, if there exist matrices $X_\varepsilon = X_\varepsilon^T$, $Y_\varepsilon = Y_\varepsilon^T$, $\mathcal{B}_i(\varepsilon)$ and $\mathcal{C}_i(\varepsilon)$, $i = 1, 2, \cdots, r$, satisfying the following ε-dependent linear matrix inequalities:*

$$\begin{bmatrix} X_\varepsilon & I \\ I & Y_\varepsilon \end{bmatrix} > 0 \tag{8.25}$$

$$X_\varepsilon > 0 \tag{8.26}$$

$$Y_\varepsilon > 0 \tag{8.27}$$

$$\Psi_{11_{ii}}(\varepsilon) < 0, \quad i = 1, 2, \cdots, r \tag{8.28}$$

$$\Psi_{22_{ii}}(\varepsilon) < 0, \quad i = 1, 2, \cdots, r \tag{8.29}$$

$$\Psi_{11_{ij}}(\varepsilon) + \Psi_{11_{ji}}(\varepsilon) < 0, \quad i < j \leq r \tag{8.30}$$

$$\Psi_{22_{ij}}(\varepsilon) + \Psi_{22_{ji}}(\varepsilon) < 0, \quad i < j \leq r \tag{8.31}$$

where

$$\Psi_{11_{ij}}(\varepsilon) = \begin{pmatrix} \begin{pmatrix} E_\varepsilon^{-1}A_iY_\varepsilon + Y_\varepsilon A_i^T E_\varepsilon^{-1} + E_\varepsilon^{-1}B_{2_i}\mathcal{C}_j(\varepsilon)E_\varepsilon^{-1} \\ +E_\varepsilon^{-1}\mathcal{C}_i^T(\varepsilon)B_{2_j}^T E_\varepsilon^{-1} + \gamma^{-2}E_\varepsilon^{-1}\tilde{B}_{1_i}\tilde{B}_{1_j}^T E_\varepsilon^{-1} \end{pmatrix} & (*)^T \\ \left[Y_\varepsilon \tilde{C}_{1_i}^T + E_\varepsilon^{-1}\mathcal{C}_i^T(\varepsilon)\tilde{D}_{12_j}^T\right]^T & -I \end{pmatrix} \quad (8.32)$$

$$\Psi_{22_{ij}}(\varepsilon) = \begin{pmatrix} \begin{pmatrix} A_i^T E_\varepsilon^{-1}X_\varepsilon + X_\varepsilon E_\varepsilon^{-1}A_i + \mathcal{B}_i(\varepsilon)C_{2_j} \\ +C_{2_i}^T\mathcal{B}_j^T(\varepsilon) + \tilde{C}_{1_i}^T\tilde{C}_{1_j} \end{pmatrix} & (*)^T \\ \left[X_\varepsilon E_\varepsilon^{-1}\tilde{B}_{1_i} + \mathcal{B}_i(\varepsilon)\tilde{D}_{21_j}\right]^T & -\gamma^2 I \end{pmatrix} \quad (8.33)$$

with

$$\tilde{B}_{1_i} = \left[\delta I \; I \; \delta I \; 0 \; B_{1_i} \; 0\right], \quad \tilde{C}_{1_i} = \left[\tfrac{\gamma\rho}{\delta}H_{1_i}^T \; 0 \; \tfrac{\gamma\rho}{\delta}H_{5_i}^T \; \sqrt{2}\lambda\rho H_{4_i}^T \; \sqrt{2}\lambda C_{1_i}^T\right]^T,$$

$$\tilde{D}_{12_i} = \left[0 \; \tfrac{\gamma\rho}{\delta}H_{3_i}^T \; 0 \; \sqrt{2}\lambda\rho H_{6_i}^T \; \sqrt{2}\lambda D_{12_i}^T\right]^T, \quad \tilde{D}_{21_i} = \left[0 \; 0 \; 0 \; \delta I \; D_{21_i} \; I\right]$$

$$\text{and } \lambda = \left(1 + \rho^2 \sum_{i=1}^{r}\sum_{j=1}^{r}\left[\|H_{2_i}^T H_{2_j}\| + \|H_{7_i}^T H_{7_j}\|\right]\right)^{\frac{1}{2}},$$

then the system (8.1) has the prescribed $\mathcal{H}_\infty$ performance $\gamma > 0$. Furthermore, a suitable controller is of the form (8.24) with

$$\begin{aligned} \hat{A}_{ij}(\varepsilon) &= E_\varepsilon\left[Y_\varepsilon^{-1} - X_\varepsilon\right]^{-1}\mathcal{M}_{ij}(\varepsilon)Y_\varepsilon^{-1} \\ \hat{B}_i &= E_\varepsilon\left[Y_\varepsilon^{-1} - X_\varepsilon\right]^{-1}\mathcal{B}_i(\varepsilon) \\ \hat{C}_i &= \mathcal{C}_i(\varepsilon)E_\varepsilon^{-1}Y_\varepsilon^{-1} \end{aligned} \quad (8.34)$$

where

$$\begin{aligned} \mathcal{M}_{ij}(\varepsilon) = &-A_i^T E_\varepsilon^{-1} - X_\varepsilon E_\varepsilon^{-1}A_iY_\varepsilon - X_\varepsilon E_\varepsilon^{-1}B_{2_i}\hat{C}_jY_\varepsilon \\ &-[Y_\varepsilon^{-1} - X_\varepsilon]E_\varepsilon^{-1}\hat{B}_iC_{2_j}Y_\varepsilon - \tilde{C}_{1_i}^T\left[\tilde{C}_{1_j}Y_\varepsilon + \tilde{D}_{12_j}\hat{C}_jY_\varepsilon\right] \\ &-\gamma^{-2}\left\{X_\varepsilon E_\varepsilon^{-1}\tilde{B}_{1_i} + [Y_\varepsilon^{-1} - X_\varepsilon]E_\varepsilon^{-1}\hat{B}_i\tilde{D}_{21_i}\right\}\tilde{B}_{1_j}^T E_\varepsilon^{-1}. \end{aligned} \quad (8.35)$$

Proof: The proof can be carried out by the same technique used in Lemma 3.1 and Theorem 2. ∎

Remark 6. The LMIs given in Lemma 6 may become ill-conditioned when ε is sufficiently small, which is always the case for the SPS. In general, these ill-conditioned LMIs are very difficult to solve. Thus, to alleviate these ill-conditioned LMIs, we have the following ε-independent well-posed LMI-based sufficient conditions for the UFSPS to obtain the prescribed $\mathcal{H}_\infty$ performance. □

Theorem 12. *Consider the system (8.1). Given a prescribed $\mathcal{H}_\infty$ performance $\gamma > 0$ and a positive constant δ, if there exist matrices X_0, Y_0, $\mathcal{B}_{0_i}$ and $\mathcal{C}_{0_i}$, $i = 1, 2, \cdots, r$, satisfying the following ε-independent linear matrix inequalities:*

$$\begin{bmatrix} X_0 E + DX_0 & I \\ I & Y_0 E + DY_0 \end{bmatrix} > 0 \tag{8.36}$$

$$EX_0^T = X_0 E, \;\; X_0^T D = DX_0, \;\; X_0 E + DX_0 > 0 \tag{8.37}$$

$$EY_0^T = Y_0 E, \;\; Y_0^T D = DY_0, \;\; Y_0 E + DY_0 > 0 \tag{8.38}$$

$$\Psi_{11_{ii}} < 0, \quad i = 1, 2, \cdots, r \tag{8.39}$$

$$\Psi_{22_{ii}} < 0, \quad i = 1, 2, \cdots, r \tag{8.40}$$

$$\Psi_{11_{ij}} + \Psi_{11_{ji}} < 0, \quad i < j \le r \tag{8.41}$$

$$\Psi_{22_{ij}} + \Psi_{22_{ji}} < 0, \quad i < j \le r \tag{8.42}$$

where $E = \begin{pmatrix} I & 0 \\ 0 & 0 \end{pmatrix}$, $D = \begin{pmatrix} 0 & 0 \\ 0 & I \end{pmatrix}$,

$$\Psi_{11_{ij}} = \begin{pmatrix} A_i Y_0^T + Y_0 A_i^T + B_{2_i}\mathcal{C}_{0_j} + \mathcal{C}_{0_i}^T B_{2_j}^T + \gamma^{-2}\tilde{B}_{1_i}\tilde{B}_{1_j}^T & (*)^T \\ \left[Y_0 \tilde{C}_{1_i}^T + \mathcal{C}_{0_i}^T \tilde{D}_{12_j}^T\right]^T & -I \end{pmatrix} \tag{8.43}$$

$$\Psi_{22_{ij}} = \begin{pmatrix} A_i^T X_0^T + X_0 A_i + \mathcal{B}_{0_i} C_{2_j} + C_{2_i}^T \mathcal{B}_{0_j}^T + \tilde{C}_{1_i}^T \tilde{C}_{1_j} & (*)^T \\ \left[X_0 \tilde{B}_{1_i} + \mathcal{B}_{0_i}\tilde{D}_{21_j}\right]^T & -\gamma^2 I \end{pmatrix} \tag{8.44}$$

with

$$\tilde{B}_{1_i} = \left[\delta I \; I \; \delta I \; 0 \; B_{1_i} \; 0\right], \quad \tilde{C}_{1_i} = \left[\tfrac{\gamma\rho}{\delta} H_{1_i}^T \; 0 \; \tfrac{\gamma\rho}{\delta} H_{5_i}^T \; \sqrt{2}\lambda\rho H_{4_i}^T \; \sqrt{2}\lambda C_{1_i}^T\right]^T$$

$$\tilde{D}_{12_i} = \left[0 \; \tfrac{\gamma\rho}{\delta} H_{3_i}^T \; 0 \; \sqrt{2}\lambda\rho H_{6_i}^T \; \sqrt{2}\lambda D_{12_i}^T\right]^T, \quad \tilde{D}_{21_i} = \left[0 \; 0 \; 0 \; \delta I \; D_{21_i} \; I\right]$$

$$\text{and } \lambda = \left(1 + \rho^2 \sum_{i=1}^{r}\sum_{j=1}^{r}\left[\|H_{2_i}^T H_{2_j}\| + \|H_{7_i}^T H_{7_j}\|\right]\right)^{\frac{1}{2}},$$

then there exists a sufficiently small $\hat{\varepsilon} > 0$ such that for $\varepsilon \in (0, \hat{\varepsilon}]$, the prescribed $\mathcal{H}_\infty$ performance $\gamma > 0$ is guaranteed. Furthermore, a suitable controller is of the form (8.24) with

$$\begin{aligned} \hat{A}_{ij}(\varepsilon) &= \left[Y_\varepsilon^{-1} - X_\varepsilon\right]^{-1} \mathcal{M}_{0_{ij}}(\varepsilon) Y_\varepsilon^{-1} \\ \hat{B}_i &= \left[Y_0^{-1} - X_0\right]^{-1} \mathcal{B}_{0_i} \\ \hat{C}_i &= \mathcal{C}_{0_i} Y_0^{-1} \end{aligned} \tag{8.45}$$

where

$$\mathcal{M}_{0_{ij}}(\varepsilon) = -A_i^T - X_\varepsilon A_i Y_\varepsilon - X_\varepsilon B_{2_i}\hat{C}_j Y_\varepsilon - \left[Y_\varepsilon^{-1} - X_\varepsilon\right]\hat{B}_i C_{2_j} Y_\varepsilon$$
$$-\tilde{C}_{1_i}^T\left[\tilde{C}_{1_j} Y_\varepsilon + \tilde{D}_{12_j}\hat{C}_j Y_\varepsilon\right] - \gamma^{-2}\left\{X_\varepsilon \tilde{B}_{1_i} + \left[Y_\varepsilon^{-1} - X_\varepsilon\right]\hat{B}_i\tilde{D}_{21_i}\right\}\tilde{B}_{1_j}^T \quad (8.46)$$
$$X_\varepsilon = \left\{X_0 + \varepsilon\tilde{X}\right\}E_\varepsilon \ \textit{and}\ Y_\varepsilon^{-1} = \left\{Y_0^{-1} + \varepsilon N_\varepsilon\right\}E_\varepsilon \quad (8.47)$$

with $\tilde{X} = D\left(X_0^T - X_0\right)$ *and* $N_\varepsilon = D\left((Y_0^{-1})^T - Y_0^{-1}\right)$.

Proof: See Appendix. ■

8.3.2 Case II–$\nu(t)$ is unavailable for feedback

Now, the premise variable of the fuzzy model $\nu(t)$ is unavailable for feedback which implies μ_i is unavailable for feedback. Hence, we cannot select our controller which depends on μ_i. Thus, we select our controller as follows:

$$\begin{aligned} E_\varepsilon\dot{\hat{x}}(t) &= \textstyle\sum_{i=1}^r\sum_{j=1}^r \hat{\mu}_i\hat{\mu}_j\left[\hat{A}_{ij}(\varepsilon)\hat{x}(t) + \hat{B}_i y(t)\right] \\ u(t) &= \textstyle\sum_{i=1}^r \hat{\mu}_i\hat{C}_i\hat{x}(t) \end{aligned} \quad (8.48)$$

where $\hat{\mu}_i$ depends on the premise variable of the controller which is different from μ_i.

Let us re-express the system (8.1) in terms of $\hat{\mu}_i$, thus the plant's premise variable becomes the same as the controller's premise variable. By doing so, the result given in the previous case can then be applied here. Note that it can be done by using the same technique as in Subsection 3.3.2. After some manipulation, we get

$$\begin{aligned} E_\varepsilon\dot{x}(t) &= \textstyle\sum_{i=1}^r \hat{\mu}_i\Big[[A_i + \Delta\bar{A}_i]x(t) + [B_{1_i} + \Delta\bar{B}_{1_i}]w(t) \\ &\qquad +[B_{2_i} + \Delta\bar{B}_{2_i}]\Big]u(t), \ \ x(0) = 0 \\ z(t) &= \textstyle\sum_{i=1}^r \hat{\mu}_i\Big[[C_{1_i} + \Delta\bar{C}_{1_i}]x(t) + [D_{12_i} + \Delta\bar{D}_{12_i}]u(t)\Big] \\ y(t) &= \textstyle\sum_{i=1}^r \hat{\mu}_i\Big[[C_{2_i} + \Delta\bar{C}_{2_i}]x(t) + [D_{21_i} + \Delta\bar{D}_{21_i}]w(t)\Big] \end{aligned} \quad (8.49)$$

where

$$\Delta\bar{A}_i = \bar{F}(x,\hat{x},t)\bar{H}_{1_i},\ \ \Delta\bar{B}_{1_i} = \bar{F}(x,\hat{x},t)\bar{H}_{2_i},\ \ \Delta\bar{B}_{2_i} = \bar{F}(x,\hat{x},t)\bar{H}_{3_i},$$
$$\Delta\bar{C}_{1_i} = \bar{F}(x,\hat{x},t)\bar{H}_{4_i},\ \ \Delta\bar{C}_{2_i} = \bar{F}(x,\hat{x},t)\bar{H}_{5_i},\ \ \Delta\bar{D}_{12_i} = \bar{F}(x,\hat{x}(t),t)\bar{H}_{6_i}$$
$$\text{and}\ \ \Delta\bar{D}_{21_i} = \bar{F}(x,\hat{x},t)\bar{H}_{7_i}$$

with

$$\bar{H}_{1_i} = \left[H_{1_i}^T\ A_1^T\ \cdots A_r^T\ H_{1_1}^T \cdots H_{1_r}^T\right]^T,\ \bar{H}_{2_i} = \left[H_{2_i}^T\ B_{1_1}^T \cdots B_{1_r}^T\ H_{2_1}^T \cdots H_{2_r}^T\right]^T,$$

$\bar{H}_{3_i} = \left[H_{3_i}^T \; B_{2_1}^T \cdots B_{2_r}^T \; H_{3_1}^T \cdots H_{3_r}^T\right]^T$, $\bar{H}_{4_i} = \left[H_{4_i}^T \; C_{1_1}^T \cdots C_{1_r}^T \; H_{4_1}^T \cdots H_{4_r}^T\right]^T$,

$\bar{H}_{5_i} = \left[H_{5_i}^T \; C_{2_1}^T \cdots C_{2_r}^T \; H_{5_1}^T \cdots H_{5_r}^T\right]^T$, $\bar{H}_{6_i} = \left[H_{6_i}^T \; D_{12_1}^T \cdots D_{12_r}^T \; H_{6_1}^T \cdots H_{6_r}^T\right]^T$

$$\bar{H}_{7_i} = \left[H_{7_i}^T \; D_{21_1}^T \cdots \; D_{21_r}^T \; H_{7_1}^T \cdots H_{7_r}^T\right]^T \text{ and}$$

$\bar{F}(x(t), \hat{x}(t), t) = \Big[F(x(t), t) \; (\mu_1 - \hat{\mu}_1) \; \cdots \; (\mu_r - \hat{\mu}_r) \; F(x(t), t)(\mu_1 - \hat{\mu}_1) \cdots$ $F(x(t), t)(\mu_r - \hat{\mu}_r)\Big]$. Note that $\|\bar{F}(x(t), \hat{x}(t), t)\| \leq \bar{\rho}$ where $\bar{\rho} = \{3\rho^2 + 2\}^{\frac{1}{2}}$. $\bar{\rho}$ is derived by utilizing the concept of vector norm in the basic system control theory and the fact that $\mu_i \geq 0$, $\hat{\mu}_i \geq 0$, $\sum_{i=1}^r \mu_i = 1$ and $\sum_{i=1}^r \hat{\mu}_i = 1$.

Now, the premise variable of the system is the same as the premise variable of the controller, thus we can apply the results given in Case I.

Theorem 13. *Consider the system (8.1). Given a prescribed $\mathcal{H}_\infty$ performance $\gamma > 0$ and a positive constant δ, if there exist matrices X_0, Y_0, $\mathcal{B}_{0_i}$ and $\mathcal{C}_{0_i}$, $i = 1, 2, \cdots, r$, satisfying the following ε-independent linear matrix inequalities:*

$$\begin{bmatrix} X_0 E + D X_0 & I \\ I & Y_0 E + D Y_0 \end{bmatrix} > 0 \tag{8.50}$$

$$E X_0^T = X_0 E, \;\; X_0^T D = D X_0, \;\; X_0 E + D X_0 > 0 \tag{8.51}$$

$$E Y_0^T = Y_0 E, \;\; Y_0^T D = D Y_0, \;\; Y_0 E + D Y_0 > 0 \tag{8.52}$$

$$\Psi 11_{ii} < 0, \quad i = 1, 2, \cdots, r \tag{8.53}$$

$$\Psi 22_{ii} < 0, \quad i = 1, 2, \cdots, r \tag{8.54}$$

$$\Psi 11_{ij} + \Psi 11_{ji} < 0, \quad i < j \leq r \tag{8.55}$$

$$\Psi 22_{ij} + \Psi 22_{ji} < 0, \quad i < j \leq r \tag{8.56}$$

where $E = \begin{pmatrix} I & 0 \\ 0 & 0 \end{pmatrix}$, $D = \begin{pmatrix} 0 & 0 \\ 0 & I \end{pmatrix}$,

$$\Psi 11_{ij} = \begin{pmatrix} A_i Y_0^T + Y_0 A_i^T + B_{2_i} \mathcal{C}_{0_j} + \mathcal{C}_{0_i}^T B_{2_j}^T + \gamma^{-2} \tilde{\bar{B}}_{1_i} \tilde{\bar{B}}_{1_j}^T & (*)^T \\ \left[Y_0 \tilde{\bar{C}}_{1_i}^T + \mathcal{C}_{0_i}^T \tilde{\bar{D}}_{12_j}^T\right]^T & -I \end{pmatrix} \tag{8.57}$$

$$\Psi 22_{ij} = \begin{pmatrix} A_i^T X_0^T + X_0 A_i + \mathcal{B}_{0_i} C_{2_j} + C_{2_i}^T \mathcal{B}_{0_j}^T + \tilde{\bar{C}}_{1_i}^T \tilde{\bar{C}}_{1_j} & (*)^T \\ \left[X_0 \tilde{\bar{B}}_{1_i} + \mathcal{B}_{0_i} \tilde{\bar{D}}_{21_j}\right]^T & -\gamma^2 I \end{pmatrix} \tag{8.58}$$

with

$$\tilde{\bar{B}}_{1_i} = \left[\delta I \; I \; \delta I \; 0 \; B_{1_i} \; 0\right], \;\; \tilde{\bar{C}}_{1_i} = \left[\tfrac{\gamma\bar{\rho}}{\delta} \bar{H}_{1_i}^T \; 0 \; \tfrac{\gamma\bar{\rho}}{\delta} \bar{H}_{5_i}^T \; \sqrt{2\lambda}\bar{\rho} \bar{H}_{4_i}^T \; \sqrt{2\lambda} C_{1_i}^T\right]^T,$$

$$\tilde{\bar{D}}_{12_i} = \left[0 \; \tfrac{\gamma\bar{\rho}}{\delta} \bar{H}_{3_i}^T \; 0 \; \sqrt{2\lambda}\bar{\rho} \bar{H}_{6_i}^T \; \sqrt{2\lambda} D_{12_i}^T\right]^T, \;\; \tilde{\bar{D}}_{21_i} = \left[0 \; 0 \; 0 \; \delta I \; D_{21_i} \; I\right]$$

$$\textit{and } \bar{\lambda} = \left(1 + \bar{\rho}^2 \sum_{i=1}^{r} \sum_{j=1}^{r} \left[\|\bar{H}_{2_i}^T \bar{H}_{2_j}\| + \|\bar{H}_{7_i}^T \bar{H}_{7_j}\| \right] \right)^{\frac{1}{2}},$$

then there exists a sufficiently small $\hat{\varepsilon} > 0$ such that for $\varepsilon \in (0, \hat{\varepsilon}]$, the prescribed $\mathcal{H}_\infty$ performance $\gamma > 0$ is guaranteed. Furthermore, a suitable controller is of the form (8.48) with

$$\begin{array}{ll} \hat{A}_{ij}(\varepsilon) &= \left[Y_\varepsilon^{-1} - X_\varepsilon\right]^{-1} \mathcal{M}_{0_{ij}}(\varepsilon) Y_\varepsilon^{-1} \\ \hat{B}_i &= \left[Y_0^{-1} - X_0\right]^{-1} \mathcal{B}_{0_i} \\ \hat{C}_i &= \mathcal{C}_{0_i} Y_0^{-1} \end{array} \tag{8.59}$$

where

$$\begin{aligned} \mathcal{M}_{0_{ij}}(\varepsilon) = -A_i^T - X_\varepsilon A_i Y_\varepsilon - X_\varepsilon B_{2_i} \hat{C}_j Y_\varepsilon - \left[Y_\varepsilon^{-1} - X_\varepsilon\right] \hat{B}_i C_{2_j} Y_\varepsilon \\ -\tilde{\bar{C}}_{1_i}^T \left[\tilde{\bar{C}}_{1_j} Y_\varepsilon + \tilde{\bar{D}}_{12_j} \hat{C}_j Y_\varepsilon\right] - \gamma^{-2} \left\{ X_\varepsilon \tilde{\bar{B}}_{1_i} + \left[Y_\varepsilon^{-1} - X_\varepsilon\right] \hat{B}_i \tilde{\bar{D}}_{21_i} \right\} \tilde{\bar{B}}_{1_j}^T \end{aligned} \tag{8.60}$$

$$X_\varepsilon = \left\{ X_0 + \varepsilon \tilde{X} \right\} E_\varepsilon \ \textit{ and } \ Y_\varepsilon^{-1} = \left\{ Y_0^{-1} + \varepsilon N_\varepsilon \right\} E_\varepsilon \tag{8.61}$$

with $\tilde{X} = D\left(X_0^T - X_0\right)$ *and* $N_\varepsilon = D\left((Y_0^{-1})^T - Y_0^{-1}\right)$.

Proof: Since (8.49) is of the form of (8.1), it can be shown by employing the proof for Theorem 12. ■

8.4 Example

Consider the tunnel diode circuit shown in Figure 8.1 where the tunnel diode is characterized by

$$i_D(t) = -0.2 v_D(t) - 0.01 v_D^3(t).$$

Assume that ε is a "parasitic" inductance in the network. Let $x_1(t) = v_C(t)$ be the capacitor voltage and $x_2(t) = i_L(t)$ be the inductor current. Then, the circuit shown in Figure 8.1 can be modelled by the following state equations:

$$\begin{array}{ll} C\dot{x}_1(t) &= 0.2x_1(t) + 0.01x_1^3(t) + x_2(t) \\ \varepsilon \dot{x}_2(t) &= -x_1(t) - Rx_2(t) + u(t) + 0.1w_2(t) \\ y(t) &= Jx(t) + 0.1w_1(t) \\ z(t) &= \begin{bmatrix} x_1(t) \\ x_2(t) \end{bmatrix} \end{array} \tag{8.62}$$

where $u(t)$ is the control input, $w_1(t)$ is the measurement noise, $w_2(t)$ are is the process noise which may represent un-modelled dynamics, y(t) is the measured output, $z(t)$ is the controlled output, J is the sensor matrix, $x(t) =$

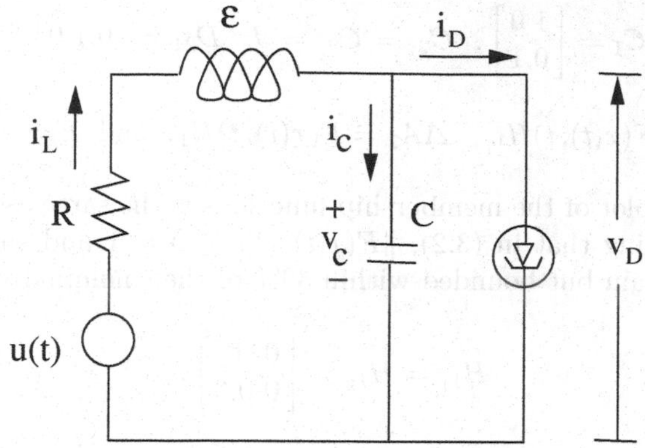

Fig. 8.1. A tunnel diode circuit.

$[x_1^T(t)\ x_2^T(t)]^T$ and $w(t) = [w_1^T(t)\ \ w_2^T(t)]^T$. Note that the variables $x_1(t)$ and $x_2(t)$ are treated as the deviation variables (variables deviate from its desired trajectories). The parameters in the circuit are given by $C = 100\ mF$ and $R = 1 \pm 0.3\%\ \Omega$, with these parameters (8.62) can be rewritten as

$$\begin{aligned} \dot{x}_1(t) &= 2x_1(t) + (0.1x_1^2(t)) \cdot x_1(t) + 10x_2(t) \\ \varepsilon\dot{x}_2(t) &= -x_1(t) - (1 \pm \Delta R)x_2(t) + u(t) + 0.1w_2(t) \\ y(t) &= Jx(t) + 0.1w_1(t) \\ z(t) &= \begin{bmatrix} x_1(t) \\ x_2(t) \end{bmatrix}. \end{aligned} \tag{8.63}$$

For the sake of simplicity, we will use as few rules as possible. Assuming that $|x_1(t)| \leq 3$, the nonlinear network system (8.63) can be approximated by the following TS fuzzy model:

Plant Rule 1: IF $x_1(t)$ is $M_1(x_1(t))$ THEN

$$\begin{aligned} E_\varepsilon \dot{x}(t) &= [A_1 + \Delta A_1]x(t) + B_1 w(t) + B_{2_1} u(t), \quad x(0) = 0, \\ z(t) &= C_1 x(t), \\ y(t) &= C_{2_1} x(t) + D_{21} w(t). \end{aligned}$$

Plant Rule 2: IF $x_1(t)$ is $M_2(x_1(t))$ THEN

$$\begin{aligned} E_\varepsilon \dot{x}(t) &= [A_2 + \Delta A_2]x(t) + B_1 w(t) + B_{2_2} u(t), \quad x(0) = 0, \\ z(t) &= C_1 x(t), \\ y(t) &= C_{2_2} x(t) + D_{21} w(t) \end{aligned}$$

where $x(t) = [x_1^T(t)\ x_2^T(t)]^T$, $w(t) = [w_1^T(t)\ w_2^T(t)]^T$,

$$A_1 = \begin{bmatrix} 2 & 10 \\ -1 & -1 \end{bmatrix}, \quad A_2 = \begin{bmatrix} 2.9 & 10 \\ -1 & -1 \end{bmatrix}, \quad B_1 = \begin{bmatrix} 0 & 0 \\ 0 & 0.1 \end{bmatrix}, \quad B_{2_1} = B_{2_2} = \begin{bmatrix} 0 \\ 1 \end{bmatrix},$$

$$C_1 = \begin{bmatrix} 1 & 0 \\ 0 & 1 \end{bmatrix}, \quad C_{2_1} = C_{2_2} = J, \quad D_{21} = \begin{bmatrix} 0.1 & 0 \end{bmatrix},$$

$$\Delta A_1 = F(x(t), t)H_{1_1}, \quad \Delta A_2 = F(x(t), t)H_{1_2} \text{ and } E_\varepsilon = \begin{bmatrix} 1 & 0 \\ 0 & \varepsilon \end{bmatrix}.$$

Note that, the plot of the membership functions is the same as in Figure 4.2. Now, by assuming that in (3.2), $\|F(x(t), t)\| \leq \rho = 1$ and since the values of R are uncertain but bounded within 30% of their nominal values given in (8.62), we have

$$H_{1_1} = H_{1_2} = \begin{bmatrix} 0 & 0 \\ 0 & 0.3 \end{bmatrix}.$$

State-feedback controller design

Employing the results given in Lemma 8.1 and the Matlab LMI solver, it is easy to realize that when $\varepsilon < 0.03$, the LMIs become ill-conditioned and the Matlab LMI solver yields the error message, "Rank Deficient". Using the LMI optimization algorithm and Theorem 11 with $\gamma = 1$ and $\delta = 1$, we obtain

$$X = \begin{bmatrix} 0.0943 & 0 \\ -0.7111 & 123.3692 \end{bmatrix},$$

$$Y_1 = \begin{bmatrix} -0.0670 & -413.6464 \end{bmatrix}, \qquad Y_2 = \begin{bmatrix} -0.0447 & -414.8722 \end{bmatrix},$$

$$K_1 = \begin{bmatrix} -25.9930 & -3.3529 \end{bmatrix}, \qquad K_2 = \begin{bmatrix} -25.8312 & -3.3629 \end{bmatrix}.$$

The resulting fuzzy controller is

$$u(t) = \sum_{j=1}^{2} \mu_j K_j x(t) \tag{8.64}$$

where

$$\mu_1 = M_1(x_1(t)) \text{ and } \mu_2 = M_2(x_1(t)).$$

Output feedback controller design

Note that by employing the results given in Lemma 8.2 and the Matlab LMI solver, it is easy to realize that when $\varepsilon < 0.03$, the LMIs become ill-conditioned and the Matlab LMI solver yields the error message, "Rank Deficient". Using the LMI optimization algorithm and Theorems 12-13 with $\varepsilon = 0.01$, $\gamma = 1$ and $\delta = 1$, we obtain the following results:

Case I-$\nu(t)$ are available for feedback

In this case, $x_1(t) = \nu(t)$ is assumed to be available for feedback; for instance, $J = [1 \;\; 0]$. This implies that μ_i is available for feedback.

$$X_0 = \begin{bmatrix} 0.3910 & 4.2648 \\ 0 & 26.0371 \end{bmatrix}, \qquad Y_0 = \begin{bmatrix} 135.2417 & -48.4620 \\ 0 & 143.3726 \end{bmatrix},$$

$$\hat{A}_{11}(\varepsilon) = \begin{bmatrix} -89.4198 & 9.2455 \\ -3.0543 & -18.6773 \end{bmatrix}, \qquad \hat{A}_{12}(\varepsilon) = \begin{bmatrix} -89.4196 & 9.2461 \\ -2.9811 & -18.5296 \end{bmatrix},$$

$$\hat{A}_{21}(\varepsilon) = \begin{bmatrix} -85.8526 & 9.2514 \\ -3.4361 & -18.6773 \end{bmatrix}, \qquad \hat{A}_{22}(\varepsilon) = \begin{bmatrix} -85.8523 & 9.2519 \\ -3.3628 & -18.5296 \end{bmatrix},$$

$$\hat{B}_1 = \begin{bmatrix} 96.6841 \\ -1.7141 \end{bmatrix}, \qquad \hat{B}_2 = \begin{bmatrix} 94.0516 \\ -1.3324 \end{bmatrix},$$

$$\hat{C}_1 = \begin{bmatrix} -46.7003 & -177.7428 \end{bmatrix}, \qquad \hat{C}_2 = \begin{bmatrix} -45.9676 & -176.2654 \end{bmatrix}.$$

The resulting fuzzy controller is

$$E_\varepsilon \dot{\hat{x}}(t) = \sum_{i=1}^{2} \sum_{j=1}^{2} \mu_i \mu_j \hat{A}_{ij}(\varepsilon) \hat{x}(t) + \sum_{i=1}^{2} \mu_i \hat{B}_i y(t)$$

$$u(t) = \sum_{i=1}^{2} \mu_i \hat{C}_i \hat{x}(t)$$

where

$$\mu_1 = M_1(x_1(t)) \text{ and } \mu_2 = M_2(x_1(t)).$$

Case II: $\nu(t)$ are unavailable for feedback

In this case, $x_1(t) = \nu(t)$ is assumed to be unavailable for feedback; for instance, $J = [0 \;\; 1]$. This implies that μ_i is unavailable for feedback.

$$X_0 = \begin{bmatrix} 0.3519 & 5.1178 \\ 0 & 43.3952 \end{bmatrix}, \qquad Y_0 = \begin{bmatrix} 225.4028 & -96.9239 \\ 0 & 143.3726 \end{bmatrix},$$

$$\hat{A}_{11}(\varepsilon) = \begin{bmatrix} -93.0660 & 9.3158 \\ -1.8731 & -19.0234 \end{bmatrix}, \qquad \hat{A}_{12}(\varepsilon) = \begin{bmatrix} -93.0657 & 9.3167 \\ -1.8291 & -18.8706 \end{bmatrix},$$

$$\hat{A}_{21}(\varepsilon) = \begin{bmatrix} -90.5479 & 9.3235 \\ -2.1021 & -19.0234 \end{bmatrix}, \qquad \hat{A}_{22}(\varepsilon) = \begin{bmatrix} -90.5476 & 9.3245 \\ -2.0581 & -18.8706 \end{bmatrix},$$

$$\hat{B}_1 = \begin{bmatrix} 100.8526 \\ -1.0284 \end{bmatrix}, \qquad \hat{B}_2 = \begin{bmatrix} 99.2575 \\ -0.7994 \end{bmatrix},$$

$$\hat{C}_1 = \begin{bmatrix} -28.0202 & -180.8999 \end{bmatrix}, \qquad \hat{C}_2 = \begin{bmatrix} -27.5806 & -179.3730 \end{bmatrix}.$$

The resulting fuzzy controller is

$$E_\varepsilon \dot{\hat{x}}(t) = \sum_{i=1}^{2}\sum_{j=1}^{2} \hat{\mu}_i \hat{\mu}_j \hat{A}_{ij}(\varepsilon)\hat{x}(t) + \sum_{i=1}^{2} \hat{\mu}_i \hat{B}_i y(t)$$

$$u(t) = \sum_{i=1}^{2} \hat{\mu}_i \hat{C}_i \hat{x}(t)$$

where

$$\hat{\mu}_1 = M_1(\hat{x}_1(t)) \text{ and } \hat{\mu}_2 = M_2(\hat{x}_1(t)).$$

Remark 7. For a sufficiently small ε, both robust fuzzy state and output feedback controllers guarantee that the $\mathcal{L}_2$-gain, γ, is less than the prescribed value. The disturbance input signal, $w(t)$, which was used during the simulation is given in Figure 8.2. The ratio of the regulated output energy to the disturbance input noise energy obtained by using the $\mathcal{H}_\infty$ fuzzy controllers with $\varepsilon = 0.01$ is depicted in Figure 8.3. After 5 seconds, the ratio of the regulated output energy to the disturbance input noise energy tends to a constant value which is about 0.032 for the state-feedback controller, and 0.10 for the output feedback controller in Case I and 0.12 in Case II. Thus, for the state-feedback controller where $\gamma = \sqrt{0.032} = 0.178$, and for output feedback controller in Case I where $\gamma = \sqrt{0.10} = 0.316$ and in Case II where $\gamma = \sqrt{0.12} = 0.346$, all are less than the prescribed value 1. Finally, Table 8.1 shows the result of the performance index γ with different values of ε. □

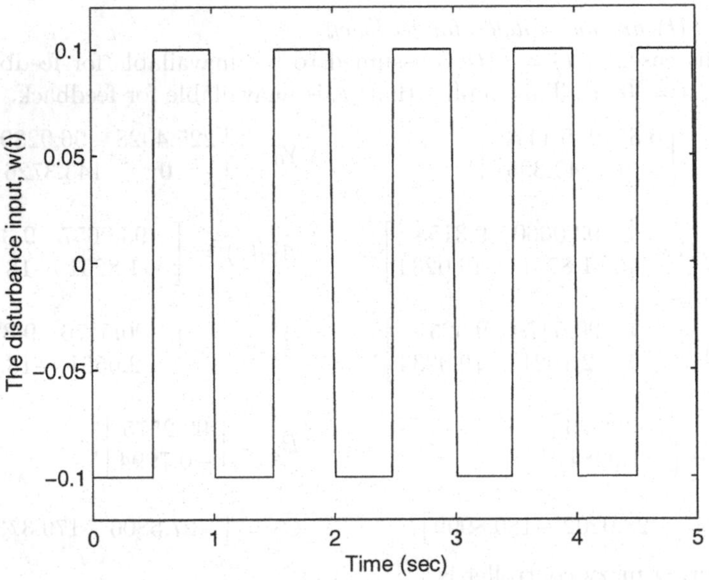

Fig. 8.2. The disturbance input noise, $w(t)$.

Table 8.1. The performance index γ of the system with different values of ε.

	The performance index γ		
ε	State-feedback	Output-feedback in Case I	Output-feedback in Case II
0.01	0.178	0.316	0.346
0.05	0.239	0.400	0.410
0.15	0.440	0.574	0.922
0.16	0.441	0.600	>1
0.28	0.500	0.989	>1
0.29	0.503	>1	>1
0.48	0.902	>1	>1
0.49	>1	>1	>1

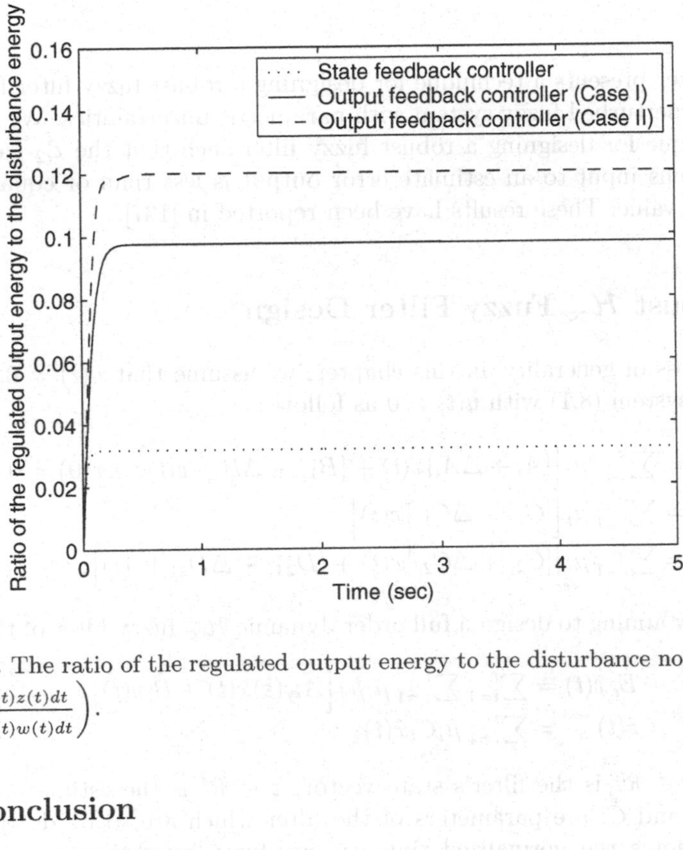

Fig. 8.3. The ratio of the regulated output energy to the disturbance noise energy: $\left(\frac{\int_0^{T_f} z^T(t)z(t)dt}{\int_0^{T_f} w^T(t)w(t)dt}\right)$.

8.5 Conclusion

This chapter has examined the problem of designing a robust fuzzy state and output feedback controller for a TS singularly perturbed fuzzy system with parametric uncertainties. Sufficient conditions for the existence of a robust fuzzy controller are derived in terms of a family of ε-independent linear matrix inequalities. A numerical simulation example has been presented to illustrate the effectiveness of the designs.

9

Robust $\mathcal{H}_\infty$ Fuzzy Filter Design for Uncertain Fuzzy Singularly Perturbed Systems

This chapter presents a technique for designing a robust fuzzy filter for a TS singularly perturbed fuzzy system with parametric uncertainties. We propose the technique for designing a robust fuzzy filter such that the $\mathcal{L}_2$-gain from an exogenous input to an estimate error output is less than or equal to the prescribed value. These results have been reported in [137].

9.1 Robust $\mathcal{H}_\infty$ Fuzzy Filter Design

Without loss of generality, in this chapter, we assume that $u(t) = 0$. Let us recall the system (8.1) with $u(t) = 0$ as follows:

$$\begin{aligned} E_\varepsilon \dot{x}(t) &= \textstyle\sum_{i=1}^r \mu_i \Big[[A_i + \Delta A_i]x(t) + [B_{1_i} + \Delta B_{1_i}]w(t) \Big], \quad x(0) = 0 \\ z(t) &= \textstyle\sum_{i=1}^r \mu_i \Big[[C_{1_i} + \Delta C_{1_i}]x(t) \Big] \\ y(t) &= \textstyle\sum_{i=1}^r \mu_i \Big[[C_{2_i} + \Delta C_{2_i}]x(t) + [D_{21_i} + \Delta D_{21_i}]w(t) \Big]. \end{aligned} \tag{9.1}$$

We are now aiming to design a full order dynamic $\mathcal{H}_\infty$ fuzzy filter of the form

$$\begin{aligned} E_\varepsilon \dot{\hat{x}}(t) &= \textstyle\sum_{i=1}^r \sum_{j=1}^r \hat{\mu}_i \hat{\mu}_j \Big[\hat{A}_{ij}(\varepsilon)\hat{x}(t) + \hat{B}_i y(t) \Big] \\ \hat{z}(t) &= \textstyle\sum_{i=1}^r \hat{\mu}_i \hat{C}_i \hat{x}(t) \end{aligned} \tag{9.2}$$

where $\hat{x}(t) \in \Re^n$ is the filter's state vector, $\hat{z} \in \Re^s$ is the estimate of $z(t)$, $\hat{A}_{ij}(\varepsilon)$, $\hat{B}_i$ and $\hat{C}_i$ are parameters of the filter which are to be determined, and $\hat{\mu}_i$ denotes the normalized time-varying fuzzy weighting functions for each rule (i.e., $\hat{\mu}_i \geq 0$ and $\sum_{i=1}^r \hat{\mu}_i = 1$), such that the inequality (4.3) holds. Clearly, in real control problems, all of the premise variables are not necessarily measurable. In this section, we then consider the designing of the robust $\mathcal{H}_\infty$ fuzzy filter into two cases as follows. Subsection 9.1.1 considers the case where the premise variable of the fuzzy model μ_i is measurable, while in Subsection 9.1.2, the premise variable is assumed to be unmeasurable.

9.1.1 Case I–$\nu(t)$ is available for feedback

The premise variable of the fuzzy model $\nu(t)$ is available for feedback which implies that μ_i is available for feedback. Thus, we can select our filter that depends on μ_i as follows [91]:

$$\begin{aligned} E_\varepsilon \dot{\hat{x}}(t) &= \sum_{i=1}^r \sum_{j=1}^r \mu_i \mu_j \left[\hat{A}_{ij}(\varepsilon) \hat{x}(t) + \hat{B}_i y(t) \right] \\ \hat{z}(t) &= \sum_{i=1}^r \mu_i \hat{C}_i \hat{x}(t). \end{aligned} \tag{9.3}$$

Before presenting our next results, the following lemma is recalled.

Lemma 7. *Consider the system (9.1). Given a prescribed $\mathcal{H}_\infty$ performance $\gamma > 0$ and a positive constant δ, if there exist matrices $X_\varepsilon = X_\varepsilon^T$, $Y_\varepsilon = Y_\varepsilon^T$, $\mathcal{B}_i(\varepsilon)$ and $\mathcal{C}_i(\varepsilon)$, $i = 1, 2, \cdots, r$, satisfying the following ε-dependent linear matrix inequalities:*

$$\begin{bmatrix} X_\varepsilon & I \\ I & Y_\varepsilon \end{bmatrix} > 0 \tag{9.4}$$

$$X_\varepsilon > 0 \tag{9.5}$$

$$Y_\varepsilon > 0 \tag{9.6}$$

$$\Psi_{11_{ii}}(\varepsilon) < 0, \quad i = 1, 2, \cdots, r \tag{9.7}$$

$$\Psi_{22_{ii}}(\varepsilon) < 0, \quad i = 1, 2, \cdots, r \tag{9.8}$$

$$\Psi_{11_{ij}}(\varepsilon) + \Psi_{11_{ji}}(\varepsilon) < 0, \quad i < j \le r \tag{9.9}$$

$$\Psi_{22_{ij}}(\varepsilon) + \Psi_{22_{ji}}(\varepsilon) < 0, \quad i < j \le r \tag{9.10}$$

where

$$\Psi_{11_{ij}}(\varepsilon) = \begin{pmatrix} E_\varepsilon^{-1} A_i Y_\varepsilon + Y_\varepsilon A_i^T E_\varepsilon^{-1} + \gamma^{-2} E_\varepsilon^{-1} \tilde{B}_{1_i} \tilde{B}_{1_j}^T E_\varepsilon^{-1} & (*)^T \\ [Y_\varepsilon \tilde{C}_{1_i}^T + E_\varepsilon^{-1} \mathcal{C}_i^T(\varepsilon) \tilde{D}_{12}^T]^T & -I \end{pmatrix} \tag{9.11}$$

$$\Psi_{22_{ij}}(\varepsilon) = \begin{pmatrix} \begin{pmatrix} A_i^T E_\varepsilon^{-1} X_\varepsilon + X_\varepsilon E_\varepsilon^{-1} A_i + \mathcal{B}_i(\varepsilon) C_{2_j} \\ + C_{2_i}^T \mathcal{B}_j^T(\varepsilon) + \tilde{C}_{1_i}^T \tilde{C}_{1_j} \end{pmatrix} & (*)^T \\ [X_\varepsilon E_\varepsilon^{-1} \tilde{B}_{1_i} + \mathcal{B}_i(\varepsilon) \tilde{D}_{21_j}]^T & -\gamma^2 I \end{pmatrix} \tag{9.12}$$

with

$$\tilde{B}_{1_i} = \left[\delta I \; I \; 0 \; B_{1_i} \; 0 \right], \quad \tilde{C}_{1_i} = \left[\tfrac{\gamma\rho}{\delta} H_{1_i}^T \; \tfrac{\gamma\rho}{\delta} H_{5_i}^T \; \sqrt{2}\lambda\rho H_{4_i}^T \; \sqrt{2}\lambda C_{1_i}^T \right]^T,$$

$$\tilde{D}_{12} = \left[0 \; 0 \; 0 \; -\sqrt{2}\lambda I \right]^T, \quad \tilde{D}_{21_i} = \left[0 \; 0 \; \delta I \; D_{21_i} \; I \right]$$

$$\text{and } \lambda = \left(1 + \rho^2 \sum_{i=1}^r \sum_{j=1}^r \left[\|H_{2_i}^T H_{2_j}\| + \|H_{7_i}^T H_{7_j}\| \right] \right)^{\frac{1}{2}},$$

then the prescribed $\mathcal{H}_\infty$ performance $\gamma > 0$ is guaranteed. Furthermore, a suitable filter is of the form (9.3) with

$$\begin{array}{ll} \hat{A}_{ij}(\varepsilon) & = E_\varepsilon\left[Y_\varepsilon^{-1} - X_\varepsilon\right]^{-1}\mathcal{M}_{ij}(\varepsilon)Y_\varepsilon^{-1} \\ \hat{B}_i & = E_\varepsilon\left[Y_\varepsilon^{-1} - X_\varepsilon\right]^{-1}\mathcal{B}_i(\varepsilon) \\ \hat{C}_i & = \mathcal{C}_i(\varepsilon)E_\varepsilon^{-1}Y_\varepsilon^{-1} \end{array} \tag{9.13}$$

where

$$\begin{aligned} \mathcal{M}_{ij}(\varepsilon) = & -A_i^T E_\varepsilon^{-1} - X_\varepsilon E_\varepsilon^{-1} A_i Y_\varepsilon - \left[Y_\varepsilon^{-1} - X_\varepsilon\right] E_\varepsilon^{-1} \hat{B}_i C_{2_j} Y_\varepsilon \\ & -\tilde{C}_{1_i}^T \left[\tilde{C}_{1_j} Y_\varepsilon + \tilde{D}_{12} \hat{C}_j Y_\varepsilon\right] \\ & -\gamma^{-2}\left\{X_\varepsilon E_\varepsilon^{-1} \tilde{B}_{1_i} + \left[Y_\varepsilon^{-1} - X_\varepsilon\right] E_\varepsilon^{-1} \hat{B}_i \tilde{D}_{21_i}\right\} \tilde{B}_{1_j}^T E_\varepsilon^{-1}. \end{aligned} \tag{9.14}$$

Proof: The proof can be carried out by the same technique used in Lemma 4.1 and Theorem 4. ∎

Remark 8. The LMIs given in Lemma 7 may become ill-conditioned when ε is sufficiently small, which is always the case for the SPS. In general, these ill-conditioned LMIs are very difficult to solve. Thus, to alleviate these ill-conditioned LMIs, we have the following ε-independent well-posed LMI-based sufficient conditions for the UFSPS to obtain the prescribed $\mathcal{H}_\infty$ performance. □

Theorem 14. *Consider the system (9.1). Given a prescribed $\mathcal{H}_\infty$ performance $\gamma > 0$ and a positive constant δ, if there exist matrices X_0, Y_0, $\mathcal{B}_{0_i}$ and $\mathcal{C}_{0_i}$, $i = 1, 2, \cdots, r$, satisfying the following ε-independent linear matrix inequalities:*

$$\begin{bmatrix} X_0 E + D X_0 & I \\ I & Y_0 E + D Y_0 \end{bmatrix} > 0 \tag{9.15}$$

$$E X_0^T = X_0 E, \quad X_0^T D = D X_0, \quad X_0 E + D X_0 > 0 \tag{9.16}$$

$$E Y_0^T = Y_0 E, \quad Y_0^T D = D Y_0, \quad Y_0 E + D Y_0 > 0 \tag{9.17}$$

$$\Psi_{11_{ii}} < 0, \quad i = 1, 2, \cdots, r \tag{9.18}$$

$$\Psi_{22_{ii}} < 0, \quad i = 1, 2, \cdots, r \tag{9.19}$$

$$\Psi_{11_{ij}} + \Psi_{11_{ji}} < 0, \quad i < j \leq r \tag{9.20}$$

$$\Psi_{22_{ij}} + \Psi_{22_{ji}} < 0, \quad i < j \leq r \tag{9.21}$$

where $E = \begin{pmatrix} I & 0 \\ 0 & 0 \end{pmatrix}$, $D = \begin{pmatrix} 0 & 0 \\ 0 & I \end{pmatrix}$,

$$\Psi_{11_{ij}} = \begin{pmatrix} A_i Y_0^T + Y_0 A_i^T + \gamma^{-2}\tilde{B}_{1_i}\tilde{B}_{1_j}^T & (*)^T \\ \left[Y_0 \tilde{C}_{1_i}^T + \mathcal{C}_{0_i}^T \tilde{D}_{12}^T\right]^T & -I \end{pmatrix} \quad (9.22)$$

$$\Psi_{22_{ij}} = \begin{pmatrix} \begin{pmatrix} A_i^T X_0^T + X_0 A_i + \mathcal{B}_{0_i} C_{2_j} \\ + C_{2_i}^T \mathcal{B}_{0_j}^T + \tilde{C}_{1_i}^T \tilde{C}_{1_j} \end{pmatrix} & (*)^T \\ \left[X_0 \tilde{B}_{1_i} + \mathcal{B}_{0_i}\tilde{D}_{21_j}\right]^T & -\gamma^2 I \end{pmatrix} \quad (9.23)$$

with

$$\tilde{B}_{1_i} = \left[\delta I \; I \; 0 \; B_{1_i} \; 0\right], \quad \tilde{C}_{1_i} = \left[\tfrac{\gamma\rho}{\delta} H_{1_i}^T \; \tfrac{\gamma\rho}{\delta} H_{5_i}^T \; \sqrt{2}\lambda\rho H_{4_i}^T \; \sqrt{2}\lambda C_{1_i}^T\right]^T,$$

$$\tilde{D}_{12} = \left[0 \; 0 \; 0 \; -\sqrt{2}\lambda I\right]^T, \quad \tilde{D}_{21_i} = \left[0 \; 0 \; \delta I \; D_{21_i} \; I\right]$$

$$\text{and } \lambda = \left(1 + \rho^2 \sum_{i=1}^{r}\sum_{j=1}^{r}\left[\|H_{2_i}^T H_{2_j}\| + \|H_{7_i}^T H_{7_j}\|\right]\right)^{\frac{1}{2}},$$

then there exists a sufficiently small $\hat{\varepsilon} > 0$ such that for $\varepsilon \in (0, \hat{\varepsilon}]$, the prescribed $\mathcal{H}_\infty$ performance $\gamma > 0$ is guaranteed. Furthermore, a suitable filter is of the form (9.3) with

$$\begin{aligned} \hat{A}_{ij}(\varepsilon) &= \left[Y_\varepsilon^{-1} - X_\varepsilon\right]^{-1}\mathcal{M}_{0_{ij}}(\varepsilon) Y_\varepsilon^{-1} \\ \hat{B}_i &= \left[Y_0^{-1} - X_0\right]^{-1}\mathcal{B}_{0_i} \\ \hat{C}_i &= \mathcal{C}_{0_i} Y_0^{-1} \end{aligned} \quad (9.24)$$

where

$$\begin{aligned} \mathcal{M}_{0_{ij}}(\varepsilon) = & -A_i^T - X_\varepsilon A_i Y_\varepsilon - \left[Y_\varepsilon^{-1} - X_\varepsilon\right]\hat{B}_i C_{2_j} Y_\varepsilon - \tilde{C}_{1_i}^T\left[\tilde{C}_{1_j} Y_\varepsilon + \tilde{D}_{12}\hat{C}_j Y_\varepsilon\right] \\ & -\gamma^{-2}\left\{X_\varepsilon \tilde{B}_{1_i} + \left[Y_\varepsilon^{-1} - X_\varepsilon\right]\hat{B}_i \tilde{D}_{21_i}\right\}\tilde{B}_{1_j}^T \end{aligned} \quad (9.25)$$

$$X_\varepsilon = \left\{X_0 + \varepsilon \tilde{X}\right\}E_\varepsilon \text{ and } Y_\varepsilon^{-1} = \left\{Y_0^{-1} + \varepsilon N_\varepsilon\right\}E_\varepsilon \quad (9.26)$$

with $\tilde{X} = D\left(X_0^T - X_0\right)$ and $N_\varepsilon = D\left((Y_0^{-1})^T - Y_0^{-1}\right)$.

Proof: The proof can be carried out by the same technique used in Theorem 12. ■

9.1.2 Case II–$\nu(t)$ is unavailable for feedback

Now, the premise variable of the fuzzy model $\nu(t)$ is unavailable for feedback which implies μ_i is unavailable for feedback. Hence, we cannot select our filter which depends on μ_i. Thus, we select our filter as follows [91]:

$$\begin{aligned} E_\varepsilon \dot{\hat{x}}(t) &= \sum_{i=1}^{r}\sum_{j=1}^{r}\hat{\mu}_i\hat{\mu}_j\left[\hat{A}_{ij}(\varepsilon)\hat{x}(t) + \hat{B}_i y(t)\right] \\ \hat{z}(t) &= \sum_{i=1}^{r}\hat{\mu}_i\hat{C}_i\hat{x}(t) \end{aligned} \quad (9.27)$$

where $\hat{\mu}_i$ depends on the premise variable of the filter which is different from μ_i.

By applying the same technique used in Subsection 3.3.2, we have the following theorem.

Theorem 15. *Consider the system (9.1). Given a prescribed $\mathcal{H}_\infty$ performance $\gamma > 0$ and a positive constant δ, if there exist matrices X_0, Y_0, $\mathcal{B}_{0_i}$ and $\mathcal{C}_{0_i}$, $i = 1, 2, \cdots, r$, satisfying the following ε-independent linear matrix inequalities:*

$$\begin{bmatrix} X_0E + DX_0 & I \\ I & Y_0E + DY_0 \end{bmatrix} > 0 \tag{9.28}$$

$$EX_0^T = X_0E, \ \ X_0^TD = DX_0, \ \ X_0E + DX_0 > 0 \tag{9.29}$$

$$EY_0^T = Y_0E, \ \ Y_0^TD = DY_0, \ \ Y_0E + DY_0 > 0 \tag{9.30}$$

$$\Psi_{11_{ii}} < 0, \quad i = 1, 2, \cdots, r \tag{9.31}$$

$$\Psi_{22_{ii}} < 0, \quad i = 1, 2, \cdots, r \tag{9.32}$$

$$\Psi_{11_{ij}} + \Psi_{11_{ji}} < 0, \quad i < j \leq r \tag{9.33}$$

$$\Psi_{22_{ij}} + \Psi_{22_{ji}} < 0, \quad i < j \leq r \tag{9.34}$$

where $E = \begin{pmatrix} I & 0 \\ 0 & 0 \end{pmatrix}$, $D = \begin{pmatrix} 0 & 0 \\ 0 & I \end{pmatrix}$,

$$\Psi_{11_{ij}} = \begin{pmatrix} A_iY_0^T + Y_0A_i^T + \gamma^{-2}\tilde{\bar{B}}_{1_i}\tilde{\bar{B}}_{1_j}^T & (*)^T \\ \left[Y_0\tilde{\bar{C}}_{1_i}^T + \mathcal{C}_{0_i}^T\tilde{\bar{D}}_{12}^T\right]^T & -I \end{pmatrix} \tag{9.35}$$

$$\Psi_{22_{ij}} = \begin{pmatrix} \begin{pmatrix} A_i^TX_0^T + X_0A_i + \mathcal{B}_{0_i}C_{2_j} \\ +C_{2_i}^T\mathcal{B}_{0_j}^T + \tilde{\bar{C}}_{1_i}^T\tilde{\bar{C}}_{1_j} \end{pmatrix} & (*)^T \\ \left[X_0\tilde{\bar{B}}_{1_i} + \mathcal{B}_{0_i}\tilde{\bar{D}}_{21_j}\right]^T & -\gamma^2 I \end{pmatrix} \tag{9.36}$$

with

$$\tilde{\bar{B}}_{1_i} = \left[\delta I \ I \ 0 \ B_{1_i} \ 0\right], \ \ \tilde{\bar{C}}_{1_i} = \left[\tfrac{\gamma\bar{\rho}}{\delta}\bar{H}_{1_i}^T \ \tfrac{\gamma\bar{\rho}}{\delta}\bar{H}_{5_i}^T \ \sqrt{2}\bar{\lambda}\bar{\rho}\bar{H}_{4_i}^T \ \sqrt{2}\bar{\lambda}C_{1_i}^T\right]^T,$$

$$\tilde{\bar{D}}_{12} = \left[0 \ 0 \ 0 \ -\sqrt{2}\bar{\lambda}I\right]^T, \ \ \tilde{\bar{D}}_{21_i} = \left[0 \ 0 \ \delta I \ D_{21_i} \ I\right]$$

$$\text{and } \bar{\lambda} = \left(1 + \bar{\rho}^2\sum_{i=1}^{r}\sum_{j=1}^{r}\left[\|\bar{H}_{2_i}^T\bar{H}_{2_j}\| + \|\bar{H}_{7_i}^T\bar{H}_{7_j}\|\right]\right)^{\frac{1}{2}},$$

then there exists a sufficiently small $\hat{\varepsilon} > 0$ such that for $\varepsilon \in (0, \hat{\varepsilon}]$, the prescribed $\mathcal{H}_\infty$ performance $\gamma > 0$ is guaranteed. Furthermore, a suitable filter is of the form (9.27) with

$$\begin{array}{ll} \hat{A}_{ij}(\varepsilon) = \left[Y_\varepsilon^{-1} - X_\varepsilon\right]^{-1} \mathcal{M}_{0_{ij}}(\varepsilon) Y_\varepsilon^{-1} & \\ \hat{B}_i \quad = \left[Y_0^{-1} - X_0\right]^{-1} \mathcal{B}_{0_i} & (9.37) \\ \hat{C}_i \quad = \mathcal{C}_{0_i} Y_0^{-1} & \end{array}$$

where

$$\begin{aligned} \mathcal{M}_{0_{ij}}(\varepsilon) &= -A_i^T - X_\varepsilon A_i Y_\varepsilon - \left[Y_\varepsilon^{-1} - X_\varepsilon\right]\hat{B}_i C_{2_j} Y_\varepsilon - \tilde{\bar{C}}_{1_i}^T \left[\tilde{\bar{C}}_{1_j} Y_\varepsilon + \tilde{\bar{D}}_{12} \hat{C}_j Y_\varepsilon\right] \\ &\quad -\gamma^{-2} \left\{ X_\varepsilon \tilde{\bar{B}}_{1_i} + \left[Y_\varepsilon^{-1} - X_\varepsilon\right] \hat{B}_i \tilde{\bar{D}}_{21_i} \right\} \tilde{\bar{B}}_{1_j}^T \end{aligned} \quad (9.38)$$

$$X_\varepsilon = \left\{X_0 + \varepsilon \tilde{X}\right\} E_\varepsilon \ \ and \ \ Y_\varepsilon^{-1} = \left\{Y_0^{-1} + \varepsilon N_\varepsilon\right\} E_\varepsilon \quad (9.39)$$

with $\tilde{X} = D\Big(X_0^T - X_0\Big)$ *and* $N_\varepsilon = D\Big((Y_0^{-1})^T - Y_0^{-1}\Big)$.

Proof: It can be shown by employing the same technique used in the proof for Theorem 13. ∎

9.2 Example

Consider the tunnel diode circuit shown in Figure 4.4 where the tunnel diode is characterized by

$$i_D(t) = 0.01 v_D(t) + 0.05 v_D^3(t).$$

Assuming that the inductance, L, is the parasitic parameter and letting $x_1(t) = v_C(t)$ and $x_2(t) = i_L(t)$ as the state variables, we have

$$\begin{aligned} C\dot{x}_1(t) &= -0.01 x_1(t) - 0.05 x_1^3(t) + x_2(t) \\ L\dot{x}_2(t) &= -x_1(t) - R x_2(t) + 0.1 w_2(t) \\ y(t) &= Jx(t) + 0.1 w_1(t) \\ z(t) &= \begin{bmatrix} x_1(t) \\ x_2(t) \end{bmatrix} \end{aligned} \quad (9.40)$$

where $w(t)$ is the disturbance noise input, $y(t)$ is the measurement output, $z(t)$ is the state to be estimated and J is the sensor matrix. Note that the variables $x_1(t)$ and $x_2(t)$ are treated as the deviation variables (variables deviate from the desired trajectories). The parameters of the circuit are $C = 100\ mF$, $R = 10 \pm 10\%\ \Omega$ and $L = \varepsilon\ H$. With these parameters (9.40) can be rewritten as

$$\begin{aligned} \dot{x}_1(t) &= -0.1 x_1(t) + 0.5 x_1^3(t) + 10 x_2(t) \\ \varepsilon \dot{x}_2(t) &= -x_1(t) - (10 + \Delta R) x_2(t) + 0.1 w_2(t) \\ y(t) &= Jx(t) + 0.1 w_1(t) \\ z(t) &= \begin{bmatrix} x_1(t) \\ x_2(t) \end{bmatrix}. \end{aligned} \quad (9.41)$$

For the sake of simplicity, we will use as few rules as possible. Assuming that $|x_1(t)| \leq 3$, the nonlinear network system (9.41) can be approximated by the following TS fuzzy model:
Plant Rule 1: IF $x_1(t)$ is $M_1(x_1(t))$ THEN

$$\begin{aligned} E_\varepsilon \dot{x}(t) &= [A_1 + \Delta A_1]x(t) + B_{1_1}w(t), \quad x(0) = 0, \\ z(t) &= C_{1_1}x(t), \\ y(t) &= C_{2_1}x(t) + D_{21_1}w(t). \end{aligned}$$

Plant Rule 2: IF $x_1(t)$ is $M_2(x_1(t))$ THEN

$$\begin{aligned} E_\varepsilon \dot{x}(t) &= [A_2 + \Delta A_2]x(t) + B_{1_2}w(t), \quad x(0) = 0, \\ z(t) &= C_{1_2}x(t), \\ y(t) &= C_{2_2}x(t) + D_{21_2}w(t) \end{aligned}$$

where $x(t) = [x_1^T(t) \ \ x_2^T(t)]^T$, $w(t) = [w_1^T(t) \ \ w_2^T(t)]^T$,

$$A_1 = \begin{bmatrix} -0.1 & 10 \\ -1 & -1 \end{bmatrix}, \quad A_2 = \begin{bmatrix} -4.6 & 10 \\ -1 & -1 \end{bmatrix}, \quad B_{1_1} = B_{1_2} = \begin{bmatrix} 0 & 0 \\ 0 & 0.1 \end{bmatrix},$$

$$C_1 = \begin{bmatrix} 1 & 0 \\ 0 & 1 \end{bmatrix}, \quad C_{2_1} = C_{2_2} = J, \ D_{21} = \begin{bmatrix} 0.1 & 0 \end{bmatrix},$$

$$\Delta A_1 = F(x(t), t)H_{1_1}, \ \ \Delta A_2 = F(x(t), t)H_{1_2} \ \text{and} \ E_\varepsilon = \begin{bmatrix} 1 & 0 \\ 0 & \varepsilon \end{bmatrix}.$$

Now, by assuming that $\|F(x(t), t)\| \leq \rho = 1$ and since the values of R are uncertain but bounded within 10% of their nominal values given in (9.40), we have

$$H_{1_1} = H_{1_2} = \begin{bmatrix} 0 & 0 \\ 0 & 1 \end{bmatrix}.$$

Note that the plot of the membership function Rules 1 and 2 is the same as in Figure 4.5. By employing the results given in Lemma 9.1 and the Matlab LMI solver [138], it is easy to realize that $\varepsilon < 0.006$ for the fuzzy filter design in Case I and $\varepsilon < 0.008$ for the fuzzy filter design in Case II, the LMIs become ill-conditioned and the Matlab LMI solver yields the error message, "Rank Deficient".

Case I-$\nu(t)$ are available for feedback

In this case, $x_1(t) = \nu(t)$ is assumed to be available for feedback; for instance, $J = [1 \ \ 0]$. This implies that μ_i is available for feedback. Using the LMI optimization algorithm and Theorem 14 with $\varepsilon = 100 \ \mu$H, $\gamma = 0.6$ and $\delta = 1$, we obtain the following results:

$$X_0 = \begin{bmatrix} 0.4082 & 0.3597 \\ 0 & 1.3551 \end{bmatrix}, \qquad Y_0 = \begin{bmatrix} 8.3868 & -0.4368 \\ 0 & 1.2210 \end{bmatrix},$$

$$\hat{A}_{11}(\varepsilon) = \begin{bmatrix} -0.0674 & -0.3532 \\ -30.7181 & -4.3834 \end{bmatrix}, \qquad \hat{A}_{12}(\varepsilon) = \begin{bmatrix} -0.0674 & -0.3532 \\ -30.7181 & -4.3834 \end{bmatrix},$$

$$\hat{A}_{21}(\varepsilon) = \begin{bmatrix} -0.0928 & -0.3138 \\ -34.7355 & -3.8964 \end{bmatrix}, \qquad \hat{A}_{22}(\varepsilon) = \begin{bmatrix} -0.0928 & -0.3138 \\ -34.7355 & -3.8964 \end{bmatrix},$$

$$\hat{B}_1 = \begin{bmatrix} 1.5835 \\ 3.2008 \end{bmatrix}, \qquad \hat{B}_2 = \begin{bmatrix} 1.2567 \\ 3.8766 \end{bmatrix},$$

$$\hat{C}_1 = \begin{bmatrix} -1.7640 & -0.8190 \end{bmatrix}, \qquad \hat{C}_2 = \begin{bmatrix} 4.5977 & -0.8190 \end{bmatrix}.$$

Hence, the resulting fuzzy filter is

$$E_\varepsilon \dot{\hat{x}}(t) = \sum_{i=1}^{2}\sum_{j=1}^{2} \mu_i \mu_j \hat{A}_{ij}(\varepsilon)\hat{x}(t) + \sum_{i=1}^{2} \mu_i \hat{B}_i y(t)$$

$$\hat{z}(t) = \sum_{i=1}^{2} \mu_i \hat{C}_i \hat{x}(t)$$

where

$$\mu_1 = M_1(x_1(t)) \text{ and } \mu_2 = M_2(x_1(t)).$$

Case II: $\nu(t)$ are unavailable for feedback

In this case, $x_1(t) = \nu(t)$ is assumed to be unavailable for feedback; for instance, $J = [0 \;\; 1]$. This implies that μ_i is unavailable for feedback. Using the LMI optimization algorithm and Theorem 15 with $\varepsilon = 100$ μH, $\gamma = 0.6$ and $\delta = 1$, we obtain the following results:

$$X_0 = \begin{bmatrix} 0.2572 & 0.3039 \\ 0 & 3.1682 \end{bmatrix}, \qquad Y_0 = \begin{bmatrix} 46.9058 & -2.0992 \\ 0 & 7.1117 \end{bmatrix},$$

$$\hat{A}_{11}(\varepsilon) = \begin{bmatrix} -2.3050 & -0.4186 \\ -32.3990 & -4.4443 \end{bmatrix}, \qquad \hat{A}_{12}(\varepsilon) = \begin{bmatrix} -2.3050 & -0.4186 \\ -32.3990 & -4.4443 \end{bmatrix},$$

$$\hat{A}_{21}(\varepsilon) = \begin{bmatrix} -2.3549 & -0.3748 \\ -32.4539 & -3.9044 \end{bmatrix}, \qquad \hat{A}_{22}(\varepsilon) = \begin{bmatrix} -2.3549 & -0.3748 \\ -32.4539 & -3.9044 \end{bmatrix},$$

$$\hat{B}_1 = \begin{bmatrix} -0.3053 \\ 3.9938 \end{bmatrix}, \qquad \hat{B}_2 = \begin{bmatrix} -0.3734 \\ 5.1443 \end{bmatrix},$$

$$\hat{C}_1 = \begin{bmatrix} 4.3913 & -0.1406 \end{bmatrix}, \qquad \hat{C}_2 = \begin{bmatrix} 1.9832 & -0.1406 \end{bmatrix}.$$

The resulting fuzzy filter is

$$E_\varepsilon \dot{\hat{x}}(t) = \sum_{i=1}^{2}\sum_{j=1}^{2} \hat{\mu}_i \hat{\mu}_j \hat{A}_{ij}(\varepsilon)\hat{x}(t) + \sum_{i=1}^{2} \hat{\mu}_i \hat{B}_i y(t)$$

$$\hat{z}(t) = \sum_{i=1}^{2} \hat{\mu}_i \hat{C}_i \hat{x}(t)$$

where

$$\hat{\mu}_1 = M_1(\hat{x}_1(t)) \text{ and } \hat{\mu}_2 = M_2(\hat{x}_1(t)).$$

Remark 9. The ratios of the filter error energy to the disturbance input noise energy are depicted in Figure 9.1 when $\varepsilon = 100\ \mu H$. *The disturbance input signal,* $w(t)$, *which was used during the simulation is the rectangular signal (magnitude 0.9 and frequency 0.5 Hz). Figures 9.2(a)-9.2(b), respectively, show the responses of* $x_1(t)$ *and* $x_2(t)$ *in Cases I and II. Table 9.1 shows the performance index* γ *with different values of* ε *in Cases I and II. After* 50 *seconds, the ratio of the filter error energy to the disturbance input noise energy tends to a constant value which is about* 0.02 *in Case I and* 0.08 *in Case II. Thus, in Case I where* $\gamma = \sqrt{0.02} = 0.141$ *and in Case II where* $\gamma = \sqrt{0.08} = 0.283$, *both are less than the prescribed value* 0.6. *From Table 9.1, the maximum value*

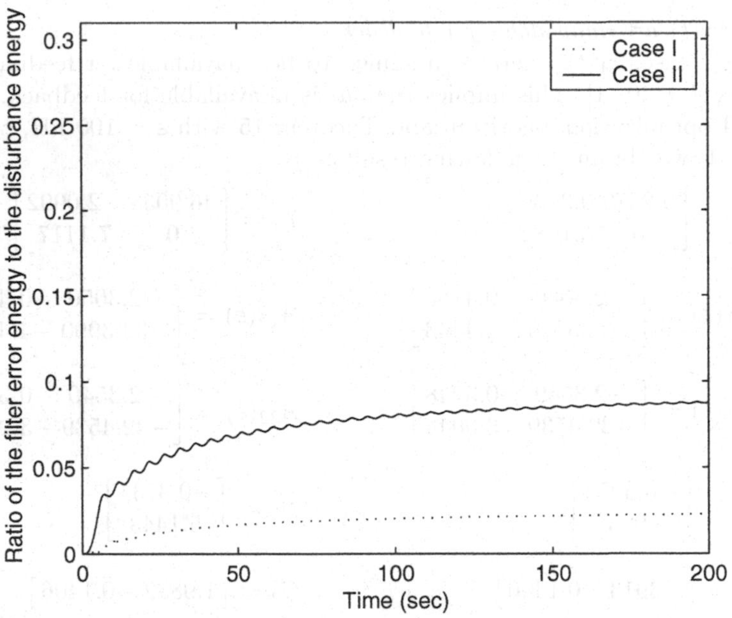

Fig. 9.1. The ratio of the filter error energy to the disturbance noise energy: $\left(\frac{\int_0^{T_f}(z(t)-\hat{z}(t))^T(z(t)-\hat{z}(t))dt}{\int_0^{T_f} w^T(t)w(t)dt}\right)$.

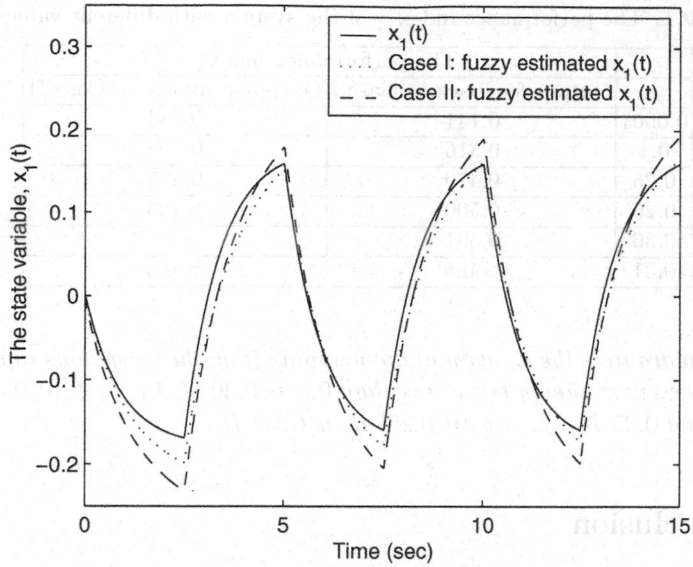

(a) The histories of $x_1(t)$

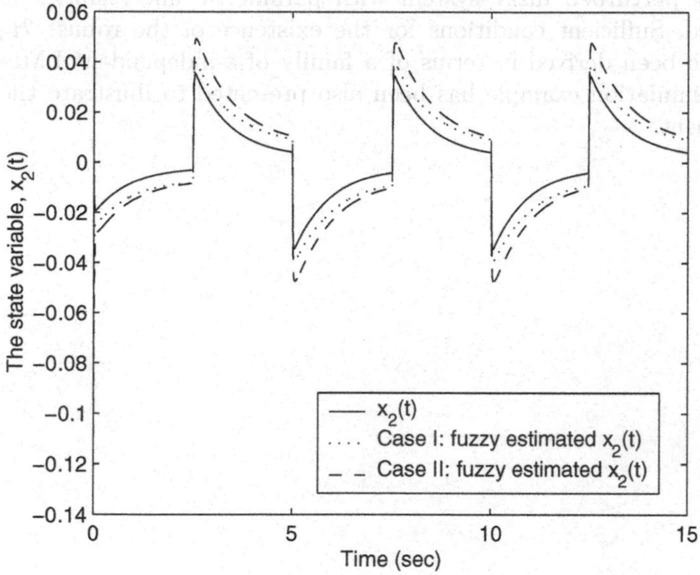

(b) The histories of $x_2(t)$

Fig. 9.2. The histories of the state variables, $x_1(t)$ and $x_2(t)$.

Table 9.1. The performance index γ of the system with different values of ε.

	The performance index γ	
ε	Output-feedback in Case I	Output-feedback in Case II
0.0001	0.141	0.283
0.1	0.316	0.509
0.25	0.479	0.596
0.26	0.500	> 0.6
0.30	0.591	> 0.6
0.31	> 0.6	> 0.6

of ε that guarantees the $\mathcal{L}_2$-gain of the mapping from the exogenous input noise to the filter error energy being less than 0.6 *is* 0.30 *H, i.e.,* $\varepsilon \in (0, 0.30]$ *H in Case I, and* 0.25 *H, i.e.,* $\varepsilon \in (0, 0.25]$ *H in Case II.* □

9.3 Conclusion

The problem of designing a robust $\mathcal{H}_\infty$ fuzzy ε-independent filter for a TS singularly perturbed fuzzy system with parametric uncertainties has been considered. Sufficient conditions for the existence of the robust $\mathcal{H}_\infty$ fuzzy filter have been derived in terms of a family of ε-independent LMIs. A numerical simulation example has been also presented to illustrate the theory development.

10

Robust $\mathcal{H}_\infty$ Fuzzy Control Design for Uncertain Fuzzy Singularly Perturbed Systems with Markovian Jumps

We propose a new technique for designing a robust fuzzy state and output feedback controller for a TS singularly perturbed fuzzy system with MJs and parametric uncertainties. The proposed robust fuzzy controller guarantees the $\mathcal{H}_\infty$ performance requirements. At the end of chapter, a numerical example is given to demonstrate the effectiveness of the proposed design procedure.

10.1 System Description

The UNSPS–MJs under consideration can be described by a TS singularly perturbed fuzzy system with MJs and parametric uncertainties as follows:

$$
\begin{aligned}
E_\varepsilon \dot{x}(t) &= \textstyle\sum_{i=1}^{r} \mu_i(\nu(t))\Big[[A_i(\eta(t)) + \Delta A_i(\eta(t))]x(t) \\
&\quad + [B_{1_i}(\eta(t)) + \Delta B_{1_i}(\eta(t))]w(t) + [B_{2_i}(\eta(t)) + \Delta B_{2_i}(\eta(t))]u(t)\Big] \\
z(t) &= \textstyle\sum_{i=1}^{r} \mu_i(\nu(t))\Big[[C_{1_i}(\eta(t)) + \Delta C_{1_i}(\eta(t))]x(t) \\
&\quad + [D_{12_i}(\eta(t)) + \Delta D_{12_i}(\eta(t))]u(t)\Big] \\
y(t) &= \textstyle\sum_{i=1}^{r} \mu_i(\nu(t))\Big[[C_{2_i}(\eta(t)) + \Delta C_{2_i}(\eta(t))]x(t) \\
&\quad + [D_{21_i}(\eta(t)) + \Delta D_{21_i}(\eta(t))]w(t)\Big]
\end{aligned}
\tag{10.1}
$$

with $x(0) = 0$, where $E_\varepsilon = \begin{bmatrix} I & 0 \\ 0 & \varepsilon I \end{bmatrix}$, $\varepsilon > 0$ is the singular perturbation parameter, $\nu(t) = \begin{bmatrix} \nu_1(t) & \cdots & \nu_\vartheta(t) \end{bmatrix}$ is the premise variable vector that may depend on states in many cases, $\mu_i(\nu(t))$ denotes the normalized time-varying fuzzy weighting functions for each rule (i.e., $\mu_i(\nu(t)) \geq 0$ and $\sum_{i=1}^{r} \mu_i(\nu(t)) = 1$), ϑ is the number of fuzzy sets, $x(t) \in \Re^n$ is the state vector, $u(t) \in \Re^m$ is the input, $w(t) \in \Re^p$ is the disturbance which belongs to $\mathcal{L}_2[0, \infty)$, $y(t) \in \Re^\ell$ is the measurement, $z(t) \in \Re^s$ is the controlled output, and the matrix functions

$A_i(\eta(t))$, $B_{1_i}(\eta(t))$, $B_{2_i}(\eta(t))$, $C_{1_i}(\eta(t))$, $C_{2_i}(\eta(t))$, $D_{12_i}(\eta(t))$, $D_{21_i}(\eta(t))$, $\Delta A_i(\eta(t))$, $\Delta B_{1_i}(\eta(t))$, $\Delta B_{2_i}(\eta(t))$, $\Delta C_{1_i}(\eta(t))$, $\Delta C_{2_i}(\eta(t))$, $\Delta D_{12_i}(\eta(t))$ and $\Delta D_{21_i}(\eta(t))$ are of appropriate dimensions. $\{\eta(t))\}$ is a continuous-time discrete-state Markov process taking values in a finite set $\mathcal{S} = \{1, 2, \cdots, s\}$ with transition probability matrix $Pr \stackrel{\Delta}{=} \{P_{\imath k}(t)\}$ given by

$$P_{\imath k}(t) = Pr(\eta(t+\Delta) = k|\eta(t) = \imath)$$
$$= \begin{cases} \lambda_{\imath k}\Delta + O(\Delta) & \text{if } \imath \neq k \\ 1 + \lambda_{\imath\imath}\Delta + O(\Delta) & \text{if } \imath = k \end{cases} \tag{10.2}$$

where $\Delta > 0$, and $\lim_{\Delta \longrightarrow 0} \frac{O(\Delta)}{\Delta} = 0$. Here $\lambda_{\imath k} \geq 0$ is the transition rate from mode $\imath$ (system operating mode) to mode k ($\imath \neq k$), and

$$\lambda_{\imath\imath} = -\sum_{k=1, k\neq \imath}^{s} \lambda_{\imath k}. \tag{10.3}$$

For the convenience of notations, we let $\mu_i \stackrel{\Delta}{=} \mu_i(\nu(t))$, $\eta = \eta(t)$, and any matrix $M(\mu, \imath) \stackrel{\Delta}{=} M(\mu, \eta = \imath)$. The matrix functions $\Delta A_i(\eta)$, $\Delta B_{1_i}(\eta)$, $\Delta B_{2_i}(\eta)$, $\Delta C_{1_i}(\eta)$, $\Delta C_{2_i}(\eta)$, $\Delta D_{12_i}(\eta)$ and $\Delta D_{21_i}(\eta)$ represent the time-varying uncertainties in the system and satisfy Assumption 5.1.

10.2 Robust $\mathcal{H}_\infty$ State-Feedback Control Design

The aim of this section is to design a robust $\mathcal{H}_\infty$ fuzzy state-feedback controller of the form

$$u(t) = \sum_{j=1}^{r} \mu_j K_j(\imath) x(t) \tag{10.4}$$

where $K_j(\imath)$ is the controller gain, such that the inequality (5.5) holds. Before presenting our next results, the following lemma is recalled.

Lemma 8. *Consider the system (10.1). Given a prescribed $\mathcal{H}_\infty$ performance $\gamma > 0$, for $\imath = 1, 2, \cdots, s$, if there exist matrices $P_\varepsilon(\imath) = P_\varepsilon^T(\imath)$, any positive constants $\delta(\imath)$ and matrices $Y_j(\imath)$, $j = 1, 2, \cdots, r$, such that the following ε-dependent linear matrix inequalities hold:*

$$P_\varepsilon(\imath) > 0 \tag{10.5}$$
$$\Psi_{ii}(\imath, \varepsilon) < 0, \quad i = 1, 2, \cdots, r \tag{10.6}$$
$$\Psi_{ij}(\imath, \varepsilon) + \Psi_{ji}(\imath, \varepsilon) < 0, \quad i < j \leq r \tag{10.7}$$

where

$$\Psi_{ij}(\imath,\varepsilon) = \begin{pmatrix} \Phi_{ij}(\imath,\varepsilon) & (*)^T & (*)^T & (*)^T \\ \mathcal{R}(\imath)\tilde{B}_{1_i}^T(\imath) & -\gamma\mathcal{R}(\imath) & (*)^T & (*)^T \\ \Upsilon_{ij}(\imath,\varepsilon) & 0 & -\gamma\mathcal{R}(\imath) & (*)^T \\ \mathcal{Z}^T(\imath,\varepsilon) & 0 & 0 & -\mathcal{P}(\imath,\varepsilon) \end{pmatrix} \tag{10.8}$$

$$\Phi_{ij}(\imath,\varepsilon) = A_i(\imath)E_\varepsilon^{-1}P_\varepsilon(\imath) + E_\varepsilon^{-1}P_\varepsilon(\imath)A_i^T(\imath) + B_{2_i}(\imath)Y_j(\imath) + Y_j^T(\imath)B_{2_i}^T(\imath) + \lambda_{\imath\imath}P_\varepsilon(\imath) \tag{10.9}$$

$$\Upsilon_{ij}(\imath,\varepsilon) = \tilde{C}_{1_i}(\imath)E_\varepsilon^{-1}P_\varepsilon(\imath) + \tilde{D}_{12_i}(\imath)Y_j(\imath) \tag{10.10}$$

$$\mathcal{R}(\imath) = diag\{\delta(\imath)I, I, \delta(\imath)I, I\} \tag{10.11}$$

$$\mathcal{Z}(\imath,\varepsilon) = \left(\sqrt{\lambda_{\imath 1}}P_\varepsilon(\imath) \cdots \sqrt{\lambda_{\imath(\imath-1)}}P_\varepsilon(\imath)\ \sqrt{\lambda_{\imath(\imath+1)}}P_\varepsilon(\imath) \cdots \sqrt{\lambda_{\imath s}}P_\varepsilon(\imath)\right) \tag{10.12}$$

$$\mathcal{P}(\imath,\varepsilon) = diag\{P_\varepsilon(1), \cdots, P_\varepsilon(\imath-1), P_\varepsilon(\imath+1), \cdots, P_\varepsilon(s)\} \tag{10.13}$$

with

$$\tilde{B}_{1_i}(\imath) = \left[I\ I\ I\ B_{1_i}(\imath) \right] \tag{10.14}$$

$$\tilde{C}_{1_i}(\imath) = \left[\gamma\rho(\imath)H_{1_i}^T(\imath)\ \sqrt{2}\aleph(\imath)\rho(\imath)H_{4_i}^T(\imath)\ 0\ \sqrt{2}\aleph(\imath)C_{1_i}^T(\imath) \right]^T \tag{10.15}$$

$$\tilde{D}_{12_i}(\imath) = \left[0\ \sqrt{2}\aleph(\imath)\rho(\imath)H_{6_i}^T(\imath)\ \gamma\rho(\imath)H_{3_i}^T(\imath)\ \sqrt{2}\aleph(\imath)D_{12_i}^T(\imath) \right]^T \tag{10.16}$$

$$\aleph(\imath) = \left(1 + \rho^2(\imath)\sum_{i=1}^{r}\sum_{j=1}^{r}\left[\| H_{2_i}^T(\imath)H_{2_j}(\imath) \| \right] \right)^{\frac{1}{2}}, \tag{10.17}$$

then the inequality (5.5) holds. Furthermore, a suitable choice of the fuzzy controller is

$$u(t) = \sum_{j=1}^{r} \mu_j K_{\varepsilon_j}(\imath)x(t) \tag{10.18}$$

where

$$K_{\varepsilon_j}(\imath) = Y_j(\imath)(P_\varepsilon(\imath))^{-1}E_\varepsilon. \tag{10.19}$$

Proof: The proof can be carried out by the same technique used in Theorem 5.1. ■

Remark 10. The LMIs given in Lemma 8 become ill-conditioned when ε is sufficiently small, which is always the case for the SPS–MJ. In general, these ill-conditioned LMIs are very difficult to solve. Thus, to alleviate these ill-conditioned LMIs, we have the following theorem which does not depend on ε. □

Theorem 16. *Consider the system (10.1). Given a prescribed $\mathcal{H}_\infty$ performance $\gamma > 0$, for $\imath = 1, 2, \cdots, s$, if there exist matrices $P(\imath)$, any positive constants $\delta(\imath)$ and matrices $Y_j(\imath)$, $j = 1, 2, \cdots, r$, such that the following ε-independent linear matrix inequalities hold:*

$$EP(\imath) = P^T(\imath)E, \quad P(\imath)D = DP^T(\imath), \quad EP(\imath) + P(\imath)D > 0 \tag{10.20}$$

$$\Psi_{ii}(\imath) < 0, \quad i = 1, 2, \cdots, r \tag{10.21}$$

$$\Psi_{ij}(\imath) + \Psi_{ji}(\imath) < 0, \quad i < j \leq r \tag{10.22}$$

where $E = \begin{pmatrix} I & 0 \\ 0 & 0 \end{pmatrix}$, $D = \begin{pmatrix} 0 & 0 \\ 0 & I \end{pmatrix}$,

$$\Psi_{ij}(\imath) = \begin{pmatrix} \Phi_{ij}(\imath) & (*)^T & (*)^T & (*)^T \\ \mathcal{R}(\imath)\tilde{B}_{1_i}^T(\imath) & -\gamma\mathcal{R}(\imath) & (*)^T & (*)^T \\ \Upsilon_{ij}(\imath) & 0 & -\gamma\mathcal{R}(\imath) & (*)^T \\ \mathcal{Z}^T(\imath) & 0 & 0 & -\mathcal{P}(\imath) \end{pmatrix} \tag{10.23}$$

$$\Phi_{ij}(\imath) = A_i(\imath)P(\imath) + P^T(\imath)A_i^T(\imath) + B_{2_i}(\imath)Y_j(\imath) + Y_j^T(\imath)B_{2_i}^T(\imath) + \lambda_{\imath\imath}\tilde{\bar{P}}(\imath) \tag{10.24}$$

$$\Upsilon_{ij}(\imath) = \tilde{C}_{1_i}(\imath)P(\imath) + \tilde{D}_{12_i}(\imath)Y_j(\imath) \tag{10.25}$$

$$\mathcal{R}(\imath) = diag\{\delta(\imath)I, I, \delta(\imath)I, I\} \tag{10.26}$$

$$\mathcal{Z}(\imath) = \left(\sqrt{\lambda_{\imath 1}}\tilde{\bar{P}}(\imath) \cdots \sqrt{\lambda_{\imath(\imath-1)}}\tilde{\bar{P}}(\imath) \ \sqrt{\lambda_{\imath(\imath+1)}}\tilde{\bar{P}}(\imath) \cdots \sqrt{\lambda_{\imath s}}\tilde{\bar{P}}(\imath)\right) \tag{10.27}$$

$$\mathcal{P}(\imath) = diag\left\{\tilde{\bar{P}}(1), \cdots, \tilde{\bar{P}}(\imath-1), \tilde{\bar{P}}(\imath+1), \cdots, \tilde{\bar{P}}(s)\right\} \tag{10.28}$$

$$\tilde{\bar{P}}(\imath) = \frac{P(\imath) + P^T(\imath)}{2} \tag{10.29}$$

with

$$\tilde{B}_{1_i}(\imath) = \left[I \ I \ I \ B_{1_i}(\imath) \right] \tag{10.30}$$

$$\tilde{C}_{1_i}(\imath) = \left[\gamma\rho(\imath)H_{1_i}^T(\imath) \ \sqrt{2}\aleph(\imath)\rho(\imath)H_{4_i}^T(\imath) \ 0 \ \sqrt{2}\aleph(\imath)C_{1_i}^T(\imath) \right]^T \tag{10.31}$$

$$\tilde{D}_{12_i}(\imath) = \left[0 \ \sqrt{2}\aleph(\imath)\rho(\imath)H_{6_i}^T(\imath) \ \gamma\rho(\imath)H_{3_i}^T(\imath) \ \sqrt{2}\aleph(\imath)D_{12_i}^T(\imath) \right]^T \tag{10.32}$$

$$\aleph(\imath) = \left(1 + \rho^2(\imath) \sum_{i=1}^{r} \sum_{j=1}^{r} \left[\| H_{2_i}^T(\imath) H_{2_j}(\imath) \| \right] \right)^{\frac{1}{2}}, \tag{10.33}$$

then there exists a sufficiently small $\hat{\varepsilon} > 0$ such that the inequality (5.5) holds for $\varepsilon \in (0, \hat{\varepsilon}]$. Furthermore, a suitable choice of the fuzzy controller is

$$u(t) = \sum_{i=1}^{r} \mu_j K_j(\imath)x(t) \tag{10.34}$$

where

$$K_j(\imath) = Y_j(\imath)(P(\imath))^{-1}. \tag{10.35}$$

Proof: Suppose there exists a matrix $P(\imath)$ such that the inequality (10.20) holds, then $P(\imath)$ is of the following form:

$$P(\imath) = \begin{pmatrix} P_1(\imath) & 0 \\ P_2^T(\imath) & P_3(\imath) \end{pmatrix} \tag{10.36}$$

with $P_1(\imath) = P_1^T(\imath) > 0$ and $P_3(\imath) = P_3^T(\imath) > 0$. Let

$$P_\varepsilon(\imath) = E_\varepsilon(P(\imath) + \varepsilon\tilde{P}(\imath)) \tag{10.37}$$

with

$$\tilde{P}(\imath) = \begin{pmatrix} 0 & P_2(\imath) \\ 0 & 0 \end{pmatrix}. \tag{10.38}$$

Substituting (10.36) and (10.38) into (10.37), we have

$$P_\varepsilon(\imath) = \begin{pmatrix} P_1(\imath) & \varepsilon P_2(\imath) \\ \varepsilon P_2^T(\imath) & \varepsilon P_3(\imath) \end{pmatrix}. \tag{10.39}$$

Clearly, $P_\varepsilon(\imath) = P_\varepsilon^T(\imath)$, and there exists a sufficiently small $\hat{\varepsilon}$ such that for $\varepsilon \in (0, \hat{\varepsilon}]$, $P_\varepsilon(\imath) > 0$. Using the matrix inversion lemma, we learn that

$$P_\varepsilon^{-1}(\imath) = \left[P^{-1}(\imath) + \varepsilon M_\varepsilon(\imath)\right] E_\varepsilon^{-1} \tag{10.40}$$

where $M_\varepsilon(\imath) = -P^{-1}(\imath)\tilde{P}(\imath)\left(I + \varepsilon P^{-1}(\imath)\tilde{P}(\imath)\right)^{-1} P^{-1}(\imath)$. Substituting (10.37) and (10.40) into (10.8), we obtain

$$\Psi_{ij}(\imath) + \psi_{ij}(\imath) \tag{10.41}$$

where the ε-independent linear matrix $\Psi_{ij}(\imath)$ is defined in (10.23) and the ε-dependent linear matrix is

$$\psi_{ij}(\imath) = \varepsilon \begin{pmatrix} \begin{pmatrix} A_i(\imath)\tilde{P}(\imath) + \tilde{P}^T(\imath)A_i^T(\imath) \\ +B_{2_i}(\imath)Y_{\varepsilon_j}(\imath) + Y_{\varepsilon_j}^T(\imath)B_{2_i}^T(\imath) + \lambda_{\imath\imath}\hat{\tilde{P}}(\imath) \end{pmatrix} & (*)^T & (*)^T & (*)^T \\ 0 & 0 & (*)^T & (*)^T \\ \left[\tilde{C}_{1_i}(\imath)\tilde{P}(\imath) + \tilde{D}_{12_i}(\imath)Y_{\varepsilon_j}(\imath)\right]^T & 0 & 0 & (*)^T \\ \mathcal{Z}_\varepsilon^T(\imath) & 0 & 0 & -\mathcal{P}_\varepsilon(\imath) \end{pmatrix}. \tag{10.42}$$

with $\mathcal{Z}_\varepsilon(\imath) = \left(\sqrt{\lambda_{\imath 1}}\hat{\tilde{P}}(\imath) \ \cdots \ \sqrt{\lambda_{\imath(\imath-1)}}\hat{\tilde{P}}(\imath) \ \sqrt{\lambda_{\imath(\imath+1)}}\hat{\tilde{P}}(\imath) \ \cdots \ \sqrt{\lambda_{\imath s}}\hat{\tilde{P}}(\imath)\right)$, $\mathcal{P}_\varepsilon(\imath) = diag\left\{\hat{\tilde{P}}(1), \cdots, \hat{\tilde{P}}(\imath-1), \hat{\tilde{P}}(\imath+1), \cdots, \hat{\tilde{P}}(s)\right\}$, $\hat{\tilde{P}}(\imath) = \frac{\tilde{P}(\imath)+\tilde{P}^T(\imath)}{2}$ and $Y_{\varepsilon_j}(\imath) = K_j(\imath)M_\varepsilon^{-1}(\imath)$. Note that the ε-dependent linear matrix tends to zero when ε approaches zero.

Employing (10.20)-(10.22) and knowing the fact that for any given negative definite matrix $\mathcal{W}$, there exists an $\varepsilon > 0$ such that $\mathcal{W} + \varepsilon I < 0$, one can show that there exists a sufficiently small $\hat{\varepsilon} > 0$ such that for $\varepsilon \in (0, \hat{\varepsilon}]$, (10.6) and (10.7) hold. Since (10.5)-(10.7) hold, using Lemma 10.1, the inequality (5.5) holds for $\varepsilon \in (0, \hat{\varepsilon}]$. ■

10.3 Robust $\mathcal{H}_\infty$ Fuzzy Output Feedback Control Design

This section aims at designing a full order dynamic $\mathcal{H}_\infty$ fuzzy output feedback controller of the form

$$\begin{array}{ll} E_\varepsilon \dot{\hat{x}}(t) = \sum_{i=1}^r \sum_{j=1}^r \hat{\mu}_i \hat{\mu}_j \left[\hat{A}_{ij}(\imath,\varepsilon)\hat{x}(t) + \hat{B}_i(\imath) y(t)\right] \\ u(t) \quad = \sum_{i=1}^r \hat{\mu}_i \hat{C}_i(\imath)\hat{x}(t) \end{array} \tag{10.43}$$

where $\hat{x}(t) \in \Re^n$ is the controller's state vector, $\hat{A}_{ij}(\imath,\varepsilon)$, $\hat{B}_i(\imath)$ and $\hat{C}_i(\imath)$ are parameters of the controller which are to be determined, and $\hat{\mu}_i$ denotes the normalized time-varying fuzzy weighting functions for each rule (i.e., $\hat{\mu}_i \geq 0$ and $\sum_{i=1}^r \hat{\mu}_i = 1$), such that the inequality (5.5) holds. Clearly, in real control problems, all of the premise variables are not necessarily measurable. Thus, we can consider the designing of the robust $\mathcal{H}_\infty$ output feedback control into two cases as follows. In Subsection 10.3.1, we consider the case where the premise variable of the fuzzy model μ_i is measurable, while in Subsection 10.3.2, the premise variable which is assumed to be unmeasurable is considered.

10.3.1 Case I–$\nu(t)$ is available for feedback

The premise variable of the fuzzy model $\nu(t)$ is available for feedback which implies that μ_i is available for feedback. Thus, we can select our controller that depends on μ_i as follows:

$$\begin{array}{ll} E_\varepsilon \dot{\hat{x}}(t) = \sum_{i=1}^r \sum_{j=1}^r \mu_i \mu_j \left[\hat{A}_{ij}(\imath,\varepsilon)\hat{x}(t) + \hat{B}_i(\imath) y(t)\right] \\ u(t) \quad = \sum_{i=1}^r \mu_i \hat{C}_i(\imath)\hat{x}(t). \end{array} \tag{10.44}$$

Before presenting our next results, the following lemma is recalled.

Lemma 9. *Consider the system (10.1). Given a prescribed $\mathcal{H}_\infty$ performance $\gamma > 0$ and any positive constants $\delta(\imath)$, for $\imath = 1, 2, \cdots, s$, if there exist matrices $X_\varepsilon(\imath) = X_\varepsilon^T(\imath)$, $Y_\varepsilon(\imath) = Y_\varepsilon^T(\imath)$, $\mathcal{B}_i(\imath,\varepsilon)$ and $\mathcal{C}_i(\imath,\varepsilon)$, $i = 1, 2, \cdots, r$, satisfying the following ε-dependent linear matrix inequalities:*

$$\begin{bmatrix} X_\varepsilon(\imath) & I \\ I & Y_\varepsilon(\imath) \end{bmatrix} > 0 \tag{10.45}$$

$$X_\varepsilon(\imath) > 0 \tag{10.46}$$

$$Y_\varepsilon(\imath) > 0 \tag{10.47}$$

$$\Psi_{11_{ii}}(\imath,\varepsilon) < 0, \quad i = 1, 2, \cdots, r \tag{10.48}$$

$$\Psi_{22_{ii}}(\imath,\varepsilon) < 0, \quad i = 1, 2, \cdots, r \tag{10.49}$$

$$\Psi_{11_{ij}}(\imath,\varepsilon) + \Psi_{11_{ji}}(\imath,\varepsilon) < 0, \quad i < j \leq r \tag{10.50}$$

$$\Psi_{22_{ij}}(\imath,\varepsilon) + \Psi_{22_{ji}}(\imath,\varepsilon) < 0, \quad i < j \leq r \tag{10.51}$$

where

$$\Psi_{11_{ij}}(\imath,\varepsilon)=\begin{pmatrix}\begin{pmatrix}E_\varepsilon^{-1}A_i(\imath)Y_\varepsilon(\imath)+Y_\varepsilon(\imath)A_i^T(\imath)E_\varepsilon^{-1}\\ +\lambda_{\imath\imath}Y_\varepsilon(\imath)E_\varepsilon^{-1}\\ +\gamma^{-2}E_\varepsilon^{-1}\tilde{B}_{1_i}(\imath)\tilde{B}_{1_j}^T(\imath)E_\varepsilon^{-1}\\ +E_\varepsilon^{-1}B_{2_i}(\imath)\mathcal{C}_j(\imath,\varepsilon)E_\varepsilon^{-1}\\ +E_\varepsilon^{-1}\mathcal{C}_i^T(\imath,\varepsilon)B_{2_j}^T(\imath)E_\varepsilon^{-1}\end{pmatrix} & (*)^T & (*)^T\\ \tilde{C}_{1_i}(\imath)Y_\varepsilon(\imath)+E_\varepsilon^{-1}\tilde{D}_{12_i}(\imath)\mathcal{C}_j(\imath,\varepsilon) & -I & (*)^T\\ \mathcal{J}^T(\imath) & 0 & -\mathcal{Y}_\varepsilon(\imath)\end{pmatrix} \quad (10.52)$$

$$\Psi_{22_{ij}}(\imath,\varepsilon)=\begin{pmatrix}\begin{pmatrix}A_i^T(\imath)E_\varepsilon^{-1}X_\varepsilon(\imath)+X_\varepsilon(\imath)E_\varepsilon^{-1}A_i(\imath)\\ +\mathcal{B}_i(\imath,\varepsilon)C_{2_j}(\imath)+C_{2_i}^T(\imath)\mathcal{B}_j^T(\imath,\varepsilon)\\ +\tilde{C}_{1_i}^T(\imath)\tilde{C}_{1_j}(\imath)+\sum_{k=1}^{s}\lambda_{\imath k}X_\varepsilon(k)E_\varepsilon^{-1}\end{pmatrix} & (*)^T\\ \left[X_\varepsilon(\imath)E_\varepsilon^{-1}\tilde{B}_{1_i}(\imath)+\mathcal{B}_i(\imath,\varepsilon)\tilde{D}_{21_j}(\imath)\right]^T & -\gamma^2 I\end{pmatrix} \quad (10.53)$$

with

$$\mathcal{J}(\imath)=\left[\sqrt{\lambda_{1\imath}}Y_\varepsilon(\imath)\ \cdots\ \sqrt{\lambda_{(i-1)\imath}}Y_\varepsilon(\imath)\ \sqrt{\lambda_{(i+1)\imath}}Y_\varepsilon(\imath)\ \cdots\ \sqrt{\lambda_{s\imath}}Y_\varepsilon(\imath)\right]$$

$$\mathcal{Y}_\varepsilon(\imath)=diag\left\{Y_\varepsilon(1),\ \cdots,\ Y_\varepsilon(\imath-1),\ Y_\varepsilon(\imath+1),\ \cdots\ ,Y_\varepsilon(s)\right\}$$

$$\tilde{B}_{1_i}(\imath)=[\delta(\imath)I\ \ I\ \ \delta(\imath)I\ \ 0\ \ B_{1_i}(\imath)\ \ 0]$$

$$\tilde{C}_{1_i}(\imath)=\left[\tfrac{\gamma\rho(\imath)}{\delta(\imath)}H_{1_i}^T(\imath)\ \ 0\ \ \tfrac{\gamma\rho(\imath)}{\delta(\imath)}H_{5_i}^T(\imath)\ \ \sqrt{2}\aleph(\imath)\rho(\imath)H_{4_i}^T(\imath)\ \ \sqrt{2}\aleph(\imath)C_{1_i}^T(\imath)\right]^T$$

$$\tilde{D}_{12_i}(\imath)=\left[0\ \ \tfrac{\gamma\rho(\imath)}{\delta(\imath)}H_{3_i}^T(\imath)\ \ 0\ \ \sqrt{2}\aleph(\imath)\rho(\imath)H_{6_i}^T(\imath)\ \ \sqrt{2}\aleph(\imath)D_{12_i}^T(\imath)\right]^T$$

$$\tilde{D}_{21_i}(\imath)=[0\ \ 0\ \ 0\ \ \delta(\imath)I\ \ D_{21_i}(\imath)\ \ I]$$

$$\aleph(\imath)=\left(1+\rho^2(\imath)\sum_{i=1}^{r}\sum_{j=1}^{r}\left[\|H_{2_i}^T(\imath)H_{2_j}(\imath)\|+\|H_{7_i}^T(\imath)H_{7_j}(\imath)\|\right]\right)^{\frac{1}{2}},$$

then the prescribed $\mathcal{H}_\infty$ performance $\gamma > 0$ is guaranteed. Furthermore, a suitable controller is of the form (10.44) with

$$\begin{aligned}\hat{A}_{ij}(\imath,\varepsilon)&=E_\varepsilon\left[Y_\varepsilon^{-1}(\imath)-X_\varepsilon(\imath)\right]^{-1}\mathcal{M}_{ij}(\imath,\varepsilon)Y_\varepsilon^{-1}(\imath)\\ \hat{B}_i(\imath)&=E_\varepsilon\left[Y_\varepsilon^{-1}(\imath)-X_\varepsilon(\imath)\right]^{-1}\mathcal{B}_i(\imath,\varepsilon)\\ \hat{C}_i(\imath)&=\mathcal{C}_i(\imath,\varepsilon)E_\varepsilon^{-1}Y_\varepsilon^{-1}(\imath)\end{aligned} \quad (10.54)$$

where

$$\begin{aligned}\mathcal{M}_{ij}(\imath,\varepsilon) = & -A_i^T(\imath)E_\varepsilon^{-1} - X_\varepsilon(\imath)E_\varepsilon^{-1}A_i(\imath)Y_\varepsilon(\imath) \\ & -\big[Y_\varepsilon^{-1}(\imath) - X_\varepsilon(\imath)\big]E_\varepsilon^{-1}\hat{B}_i(\imath)C_{2_j}(\imath)Y_\varepsilon(\imath) \\ & -X_\varepsilon E_\varepsilon^{-1}B_{2_i}(\imath)\hat{C}_j(\imath)Y_\varepsilon(\imath) - \sum_{k=1}^{s}\lambda_{\imath k}Y_\varepsilon^{-1}(k)Y_\varepsilon(\imath) \\ & -\tilde{C}_{1_i}^T(\imath)\Big[\tilde{C}_{1_j}(\imath)Y_\varepsilon(\imath) + \tilde{D}_{12_j}(\imath)\hat{C}_j(\imath)Y_\varepsilon(\imath)\Big] - \gamma^{-2}\Big\{X_\varepsilon(\imath)E_\varepsilon^{-1}\tilde{B}_{1_i}(\imath) \\ & +\big[Y_\varepsilon^{-1}(\imath) - X_\varepsilon(\imath)\big]E_\varepsilon^{-1}\hat{B}_i(\imath)\tilde{D}_{21_i}(\imath)\Big\}\tilde{B}_{1_j}^T(\imath)E_\varepsilon^{-1}. \end{aligned} \tag{10.55}$$

Proof: The proof can be carried out by the same technique used in Lemma 5.1 and Theorem 5.2. ■

Remark 11. The LMIs given in Lemma 9 may become ill-conditioned when ε is sufficiently small, which is always the case for the SPS–MJ. In general, these ill-conditioned LMIs are very difficult to solve. Thus, to alleviate these ill-conditioned LMIs, we have the following ε-independent well-posed LMI-based sufficient conditions for the UFSPS–MJ to obtain the prescribed $\mathcal{H}_\infty$ performance. □

Theorem 17. *Consider the system (10.1). Given a prescribed $\mathcal{H}_\infty$ performance $\gamma > 0$ and any positive constants $\delta(\imath)$, for $\imath = 1, 2, \cdots, s$, if there exist matrices $X_0(\imath)$, $Y_0(\imath)$, $\mathcal{B}_{0_i}(\imath)$ and $\mathcal{C}_{0_i}(\imath)$, $i = 1, 2, \cdots, r$, satisfying the following ε-independent linear matrix inequalities:*

$$\begin{bmatrix} X_0(\imath)E + DX_0(\imath) & I \\ I & Y_0(\imath)E + DY_0(\imath) \end{bmatrix} > 0 \tag{10.56}$$

$$EX_0^T(\imath) = X_0(\imath)E, \;\; X_0^T(\imath)D = DX_0(\imath), \;\; X_0(\imath)E + DX_0(\imath) > 0 \tag{10.57}$$

$$EY_0^T(\imath) = Y_0(\imath)E, \;\; Y_0^T(\imath)D = DY_0(\imath), \;\; Y_0(\imath)E + DY_0(\imath) > 0 \tag{10.58}$$

$$\Psi_{11_{ii}}(\imath) < 0, \quad i = 1, 2, \cdots, r \tag{10.59}$$

$$\Psi_{22_{ii}}(\imath) < 0, \quad i = 1, 2, \cdots, r \tag{10.60}$$

$$\Psi_{11_{ij}}(\imath) + \Psi_{11_{ji}}(\imath) < 0, \quad i < j \le r \tag{10.61}$$

$$\Psi_{22_{ij}}(\imath) + \Psi_{22_{ji}}(\imath) < 0, \quad i < j \le r \tag{10.62}$$

where $E = \begin{pmatrix} I & 0 \\ 0 & 0 \end{pmatrix}$, $D = \begin{pmatrix} 0 & 0 \\ 0 & I \end{pmatrix}$,

$$\Psi_{11_{ij}}(\imath) = \begin{pmatrix} \begin{pmatrix} A_i(\imath)Y_0^T(\imath) + Y_0(\imath)A_i^T(\imath) \\ +\lambda_{\imath\imath}\tilde{\tilde{Y}}_0(\imath) + \gamma^{-2}\tilde{B}_{1_i}(\imath)\tilde{B}_{1_j}^T(\imath) \\ +B_{2_i}(\imath)\mathcal{C}_{0_j}(\imath) + \mathcal{C}_{0_i}^T(\imath)B_{2_j}^T(\imath) \end{pmatrix} & (*)^T & (*)^T \\ \tilde{C}_{1_i}(\imath)Y_0^T(\imath) + \tilde{D}_{12_i}(\imath)\mathcal{C}_{0_j}(\imath) & -I & (*)^T \\ \mathcal{J}_0^T(\imath) & 0 & -\mathcal{Y}_0(\imath) \end{pmatrix} \tag{10.63}$$

$$\Psi_{22_{ij}}(\imath) = \begin{pmatrix} \begin{pmatrix} A_i^T(\imath)X_0^T(\imath) + X_0(\imath)A_i(\imath) \\ +\mathcal{B}_{0_i}(\imath)C_{2_j}(\imath) + C_{2_i}^T(\imath)\mathcal{B}_{0_j}^T(\imath) \\ +\tilde{C}_{1_i}^T(\imath)\tilde{C}_{1_j}(\imath) + \sum_{k=1}^{s}\lambda_{\imath k}\bar{\tilde{X}}_0(k) \end{pmatrix} & (*)^T \\ \tilde{B}_{1_i}^T(\imath)X_0^T(\imath) + \tilde{D}_{21_i}^T(\imath)\mathcal{B}_{0_j}^T(\imath) & -\gamma^2 I \end{pmatrix} \quad (10.64)$$

with $\bar{\tilde{X}}_0(k) = \frac{X_0(k)+X_0^T(k)}{2}$, $\bar{\tilde{Y}}_0(\imath) = \frac{Y_0(\imath)+Y_0^T(\imath)}{2}$,

$$\mathcal{J}_0(\imath) = \left[\sqrt{\lambda_{1\imath}}\bar{\tilde{Y}}_0(\imath) \ \cdots \ \sqrt{\lambda_{(i-1)\imath}}\bar{\tilde{Y}}_0(\imath) \ \sqrt{\lambda_{(i+1)\imath}}\bar{\tilde{Y}}_0(\imath) \ \cdots \ \sqrt{\lambda_{s\imath}}\bar{\tilde{Y}}(\imath)\right]$$

$$\mathcal{Y}_0(\imath) = diag\left\{\bar{\tilde{Y}}_0(1), \ \cdots, \ \bar{\tilde{Y}}_0(\imath-1), \ \bar{\tilde{Y}}_0(\imath+1), \ \cdots \ ,\bar{\tilde{Y}}_0(s)\right\}$$

$$\tilde{B}_{1_i}(\imath) = [\delta(\imath)I \ I \ \delta(\imath)I \ 0 \ B_{1_i}(\imath) \ 0]$$

$$\tilde{C}_{1_i}(\imath) = \left[\frac{\gamma\rho(\imath)}{\delta(\imath)}H_{1_i}^T(\imath) \ 0 \ \frac{\gamma\rho(\imath)}{\delta(\imath)}H_{5_i}^T(\imath) \ \sqrt{2}\aleph(\imath)\rho(\imath)H_{4_i}^T(\imath) \ \sqrt{2}\aleph(\imath)C_{1_i}^T(\imath)\right]^T$$

$$\tilde{D}_{12_i}(\imath) = \left[0 \ \frac{\gamma\rho(\imath)}{\delta(\imath)}H_{3_i}^T(\imath) \ 0 \ \sqrt{2}\aleph(\imath)\rho(\imath)H_{6_i}^T(\imath) \ \sqrt{2}\aleph(\imath)D_{12_i}^T(\imath)\right]^T$$

$$\tilde{D}_{21_i}(\imath) = [0 \ 0 \ 0 \ \delta(\imath)I \ D_{21_i}(\imath) \ I]$$

$$\aleph(\imath) = \left(1 + \rho^2(\imath)\sum_{i=1}^{r}\sum_{j=1}^{r}\left[\|H_{2_i}^T(\imath)H_{2_j}(\imath)\| + \|H_{7_i}^T(\imath)H_{7_j}(\imath)\|\right]\right)^{\frac{1}{2}},$$

then there exists a sufficiently small $\hat{\varepsilon} > 0$ *such that for* $\varepsilon \in (0, \hat{\varepsilon}]$, *the prescribed* $\mathcal{H}_\infty$ *performance* $\gamma > 0$ *is guaranteed. Furthermore, a suitable controller is of the form (10.44) with*

$$\begin{aligned} \hat{A}_{ij}(\imath,\varepsilon) &= \left[Y_\varepsilon^{-1}(\imath) - X_\varepsilon(\imath)\right]^{-1}\mathcal{M}_{0_{ij}}(\imath,\varepsilon)Y_\varepsilon^{-1}(\imath) \\ \hat{B}_i(\imath) &= \left[Y_0^{-1}(\imath) - X_0(\imath)\right]^{-1}\mathcal{B}_{0_i}(\imath) \\ \hat{C}_i(\imath) &= \mathcal{C}_{0_i}(\imath)Y_0^{-1}(\imath) \end{aligned} \quad (10.65)$$

where

$$\begin{aligned} \mathcal{M}_{0_{ij}}(\imath,\varepsilon) = & -A_i^T(\imath) - X_\varepsilon(\imath)A_i(\imath)Y_\varepsilon(\imath) - \left[Y_\varepsilon^{-1}(\imath) - X_\varepsilon(\imath)\right]\hat{B}_i(\imath)C_{2_j}(\imath)Y_\varepsilon(\imath) \\ & -X_\varepsilon B_{2_i}(\imath)\hat{C}_j(\imath)Y_\varepsilon(\imath) - \textstyle\sum_{k=1}^{s}\lambda_{\imath k}Y_\varepsilon^{-1}(k)Y_\varepsilon(\imath) \\ & -\tilde{C}_{1_i}^T(\imath)\left[\tilde{C}_{1_j}(\imath)Y_\varepsilon(\imath) + \tilde{D}_{12_j}(\imath)\hat{C}_j(\imath)Y_\varepsilon(\imath)\right] \\ & -\gamma^{-2}\left\{X_\varepsilon(\imath)\tilde{B}_{1_i}(\imath) + \left[Y_\varepsilon^{-1}(\imath) - X_\varepsilon(\imath)\right]\hat{B}_i(\imath)\tilde{D}_{21_i}(\imath)\right\}\tilde{B}_{1_j}^T(\imath), \end{aligned} \quad (10.66)$$

$$X_\varepsilon(\imath) = \left\{X_0(\imath) + \varepsilon\tilde{X}(\imath)\right\}E_\varepsilon \quad (10.67)$$

and

$$Y_\varepsilon^{-1}(\imath) = \left\{Y_0^{-1}(\imath) + \varepsilon N_\varepsilon(\imath)\right\}E_\varepsilon \quad (10.68)$$

with $\tilde{X}(\imath) = D\Big(X_0^T(\imath) - X_0(\imath)\Big)$ *and* $N_\varepsilon(\imath) = D\Big((Y_0^{-1}(\imath))^T - Y_0^{-1}(\imath)\Big)$.

Proof: See Appendix. ∎

10.3.2 Case II–$\nu(t)$ is unavailable for feedback

Now, the premise variable of the fuzzy model $\nu(t)$ is unavailable for feedback which implies μ_i is unavailable for feedback. Hence, we cannot select our controller which depends on μ_i. Thus, we select our controller as follows:

$$\begin{aligned} E_\varepsilon\dot{\hat{x}}(t) &= \sum_{i=1}^{r}\sum_{j=1}^{r}\hat{\mu}_i\hat{\mu}_j\Big[\hat{A}_{ij}(\imath,\varepsilon)\hat{x}(t)+\hat{B}_i(\imath)y(t)\Big] \\ u(t) &= \sum_{i=1}^{r}\hat{\mu}_i\hat{C}_i(\imath)\hat{x}(t) \end{aligned} \tag{10.69}$$

where $\hat{\mu}_i$ depends on the premise variable of the controller which is different from μ_i.

Let us re-express the system (10.1) in terms of $\hat{\mu}_i$, thus the plant's premise variable becomes the same as the controller's premise variable. By doing so, the result given in the previous case can then be applied here. Note that it can be done by using the same technique as in Subsection 3.3.2. After some manipulation, we get

$$\begin{aligned} E_\varepsilon\dot{x}(t) &= \sum_{i=1}^{r}\hat{\mu}_i\Big[[A_i(\imath)+\Delta\bar{A}_i(\imath)]x(t)+[B_{1_i}(\imath)+\Delta\bar{B}_{1_i}(\imath)]w(t) \\ &\quad +[B_{2_i}(\imath)+\Delta\bar{B}_{2_i}(\imath)]u(t)\Big],\ x(0)=0 \\ z(t) &= \sum_{i=1}^{r}\hat{\mu}_i\Big[[C_{1_i}(\imath)+\Delta\bar{C}_{1_i}(\imath)]x(t)+[D_{12_i}(\imath)+\Delta\bar{D}_{12_i}(\imath)]u(t)\Big] \\ y(t) &= \sum_{i=1}^{r}\hat{\mu}_i\Big[[C_{2_i}(\imath)+\Delta\bar{C}_{2_i}(\imath)]x(t)+[D_{21_i}(\imath)+\Delta\bar{D}_{21_i}(\imath)]w(t)\Big] \end{aligned} \tag{10.70}$$

where

$$\Delta\bar{A}_i(\imath)=\bar{F}(x(t),\hat{x}(t),\imath,t)\bar{H}_{1_i}(\imath),\quad \Delta\bar{B}_{1_i}(\imath)=\bar{F}(x(t),\hat{x}(t),\imath,t)\bar{H}_{2_i}(\imath),$$
$$\Delta\bar{B}_{2_i}(\imath)=\bar{F}(x(t),\hat{x}(t),\imath,t)\bar{H}_{3_i}(\imath),\quad \Delta\bar{C}_{1_i}(\imath)=\bar{F}(x(t),\hat{x}(t),\imath,t)\bar{H}_{4_i}(\imath),$$
$$\Delta\bar{C}_{2_i}(\imath)=\bar{F}(x(t),\hat{x}(t),\imath,t)\bar{H}_{5_i}(\imath),\quad \Delta\bar{D}_{12_i}(\imath)=\bar{F}(x(t),\hat{x}(t),\imath,t)\bar{H}_{6_i}(\imath)$$
$$\text{and}\quad \Delta\bar{D}_{21_i}(\imath)=\bar{F}(x(t),\hat{x}(t),\imath,t)\bar{H}_{7_i}(\imath)$$

with

$$\bar{H}_{1_i}(\imath)=\begin{bmatrix}H_{1_i}^T(\imath) & A_1^T(\imath) & \cdots & A_r^T(\imath) & H_{1_1}^T(\imath) & \cdots & H_{1_r}^T(\imath)\end{bmatrix}^T,$$
$$\bar{H}_{2_i}(\imath)=\begin{bmatrix}H_{2_i}^T(\imath) & B_{1_1}^T(\imath) & \cdots & B_{1_r}^T(\imath) & H_{2_1}^T(\imath) & \cdots & H_{2_r}^T(\imath)\end{bmatrix}^T,$$
$$\bar{H}_{3_i}(\imath)=\begin{bmatrix}H_{3_i}^T(\imath) & B_{2_1}^T(\imath) & \cdots & B_{2_r}^T(\imath) & H_{3_1}^T(\imath) & \cdots & H_{3_r}^T(\imath)\end{bmatrix}^T,$$
$$\bar{H}_{4_i}(\imath)=\begin{bmatrix}H_{4_i}^T(\imath) & C_{1_1}^T(\imath) & \cdots & C_{1_r}^T(\imath) & H_{4_1}^T(\imath) & \cdots & H_{4_r}^T(\imath)\end{bmatrix}^T,$$
$$\bar{H}_{5_i}(\imath)=\begin{bmatrix}H_{5_i}^T(\imath) & C_{2_1}^T(\imath) & \cdots & C_{2_r}^T(\imath) & H_{5_1}^T(\imath) & \cdots & H_{5_r}^T(\imath)\end{bmatrix}^T,$$
$$\bar{H}_{6_i}(\imath)=\begin{bmatrix}H_{6_i}^T(\imath) & D_{12_1}^T(\imath) & \cdots & D_{12_r}^T(\imath) & H_{6_1}^T(\imath) & \cdots & H_{6_r}^T(\imath)\end{bmatrix}^T,$$
$$\bar{H}_{7_i}(\imath)=\begin{bmatrix}H_{7_i}^T(\imath) & D_{21_1}^T(\imath) & \cdots & D_{21_r}^T(\imath) & H_{7_1}^T(\imath) & \cdots & H_{7_r}^T(\imath)\end{bmatrix}^T.$$

and $\bar{F}(x(t),\hat{x}(t),\imath,t) = \Big[F(x(t),\imath,t)\ (\mu_1-\hat{\mu}_1)\ \cdots\ (\mu_r-\hat{\mu}_r)\ F(x(t),\imath,t)(\mu_1-\hat{\mu}_1)\ \cdots\ F(x(t),\imath,t)(\mu_r-\hat{\mu}_r)\Big]$. Note that $\|\bar{F}(x(t),\hat{x}(t),\imath,t)\| \leq \bar{\rho}(\imath)$ where $\bar{\rho}(\imath) = \{3\rho^2(\imath)+2\}^{\frac{1}{2}}$. $\bar{\rho}(\imath)$ is derived by utilizing the concept of vector norm in the basic system control theory and the fact that $\mu_i \geq 0$, $\hat{\mu}_i \geq 0$, $\sum_{i=1}^r \mu_i = 1$ and $\sum_{i=1}^r \hat{\mu}_i = 1$.

In this new expression, the plant's premise variable is now the same as the controller's premise variable. Thus, applying Theorem 17, we have the following LMI-based sufficient conditions for this case.

Theorem 18. *Consider the system (10.1). Given a prescribed $\mathcal{H}_\infty$ performance $\gamma > 0$ and any positive constants $\delta(\imath)$, for $\imath = 1,2,\cdots,s$, if there exist matrices $X_0(\imath)$, $Y_0(\imath)$, $\mathcal{B}_{0_i}(\imath)$ and $\mathcal{C}_{0_i}(\imath)$, $i = 1,2,\cdots,r$, satisfying the following ε-independent linear matrix inequalities:*

$$\begin{bmatrix} X_0(\imath)E + DX_0(\imath) & I \\ I & Y_0(\imath)E + DY_0(\imath)\end{bmatrix} > 0 \qquad (10.71)$$

$$EX_0^T(\imath) = X_0(\imath)E,\ \ X_0^T(\imath)D = DX_0(\imath),\ \ X_0(\imath)E + DX_0(\imath) > 0 \qquad (10.72)$$

$$EY_0^T(\imath) = Y_0(\imath)E,\ \ Y_0^T(\imath)D = DY_0(\imath),\ \ Y_0(\imath)E + DY_0(\imath) > 0 \qquad (10.73)$$

$$\Psi_{11_{ii}}(\imath) < 0, \quad i = 1,2,\cdots,r \qquad (10.74)$$

$$\Psi_{22_{ii}}(\imath) < 0, \quad i = 1,2,\cdots,r \qquad (10.75)$$

$$\Psi_{11_{ij}}(\imath) + \Psi_{11_{ji}}(\imath) < 0, \quad i < j \leq r \qquad (10.76)$$

$$\Psi_{22_{ij}}(\imath) + \Psi_{22_{ji}}(\imath) < 0, \quad i < j \leq r \qquad (10.77)$$

where $E = \begin{pmatrix} I & 0 \\ 0 & 0\end{pmatrix}$, $D = \begin{pmatrix} 0 & 0 \\ 0 & I\end{pmatrix}$,

$$\Psi_{11_{ij}}(\imath) = \begin{pmatrix} \begin{pmatrix} A_i(\imath)Y_0^T(\imath) + Y_0(\imath)A_i^T(\imath) \\ +\lambda_{\imath\imath}\tilde{Y}_0(\imath) + \gamma^{-2}\tilde{\bar{B}}_{1_i}(\imath)\tilde{\bar{B}}_{1_j}^T(\imath) \\ +B_{2_i}(\imath)\mathcal{C}_{0_j}(\imath) + \mathcal{C}_{0_i}^T(\imath)B_{2_j}^T(\imath)\end{pmatrix} & (*)^T & (*)^T \\ \tilde{\bar{C}}_{1_i}(\imath)Y_0^T(\imath) + \tilde{\bar{D}}_{12_i}(\imath)\mathcal{C}_{0_j}(\imath) & -I & (*)^T \\ \mathcal{J}_0^T(\imath) & 0 & -\mathcal{Y}_0(\imath)\end{pmatrix} \qquad (10.78)$$

$$\Psi_{22_{ij}}(\imath) = \begin{pmatrix} \begin{pmatrix} A_i^T(\imath)X_0^T(\imath) + X_0(\imath)A_i(\imath) \\ +\mathcal{B}_{0_i}(\imath)C_{2_j}(\imath) + C_{2_i}^T(\imath)\mathcal{B}_{0_j}^T(\imath) \\ +\tilde{\bar{C}}_{1_i}^T(\imath)\tilde{\bar{C}}_{1_j}(\imath) + \sum_{k=1}^s \lambda_{\imath k}\tilde{X}_0(k)\end{pmatrix} & (*)^T \\ \tilde{\bar{B}}_{1_i}^T(\imath)X_0^T(\imath) + \tilde{\bar{D}}_{21_i}^T(\imath)\mathcal{B}_{0_j}^T(\imath) & -\gamma^2 I\end{pmatrix} \qquad (10.79)$$

with $\tilde{X}_0(k) = \frac{X_0(k)+X_0^T(k)}{2}$, $\tilde{Y}_0(\imath) = \frac{Y_0(\imath)+Y_0^T(\imath)}{2}$,

$$\mathcal{J}_0(\imath) = \left[\sqrt{\lambda_{1\imath}}\tilde{\bar{Y}}_0(\imath) \ \cdots \ \sqrt{\lambda_{(i-1)\imath}}\tilde{\bar{Y}}_0(\imath) \ \sqrt{\lambda_{(i+1)\imath}}\tilde{\bar{Y}}_0(\imath) \ \cdots \ \sqrt{\lambda_{s\imath}}\tilde{\bar{Y}}_0(\imath)\right]$$

$$\mathcal{Y}_0(\imath) = diag\left\{\tilde{\bar{Y}}_0(1), \ \cdots, \ \tilde{\bar{Y}}_0(\imath-1), \ \tilde{\bar{Y}}_0(\imath+1), \ \cdots \ ,\tilde{\bar{Y}}_0(s)\right\}$$

$$\tilde{\bar{B}}_{1_i}(\imath) = [\delta(\imath)I \ \ I \ \ \delta(\imath)I \ \ 0 \ \ B_{1_i}(\imath) \ \ 0]$$

$$\tilde{\bar{C}}_{1_i}(\imath) = \left[\tfrac{\gamma\bar{\rho}(\imath)}{\delta(\imath)}\bar{H}_{1_i}^T(\imath) \ \ 0 \ \ \tfrac{\gamma\bar{\rho}(\imath)}{\delta(\imath)}\bar{H}_{5_i}^T(\imath) \ \ \sqrt{2}\bar{\aleph}(\imath)\bar{\rho}(\imath)\bar{H}_{4_i}^T(\imath) \ \ \sqrt{2}\bar{\aleph}(\imath)C_{1_i}^T(\imath)\right]^T$$

$$\tilde{\bar{D}}_{12_i}(\imath) = \left[0 \ \ \tfrac{\gamma\bar{\rho}(\imath)}{\delta(\imath)}\bar{H}_{3_i}^T(\imath) \ \ 0 \ \ \sqrt{2}\bar{\aleph}(\imath)\bar{\rho}(\imath)\bar{H}_{6_i}^T(\imath) \ \ \sqrt{2}\bar{\aleph}(\imath)D_{12_i}^T(\imath)\right]^T$$

$$\tilde{\bar{D}}_{21_i}(\imath) = [0 \ \ 0 \ \ 0 \ \ \delta(\imath)I \ \ D_{21_i}(\imath) \ \ I]$$

$$\bar{\aleph}(\imath) = \left(1 + \bar{\rho}^2(\imath)\sum_{i=1}^{r}\sum_{j=1}^{r}\left[\|\bar{H}_{2_i}^T(\imath)\bar{H}_{2_j}(\imath)\| + \|\bar{H}_{7_i}^T(\imath)\bar{H}_{7_j}(\imath)\|\right]\right)^{\frac{1}{2}},$$

then there exists a sufficiently small $\hat{\varepsilon} > 0$ such that for $\varepsilon \in (0, \hat{\varepsilon}]$, the prescribed $\mathcal{H}_\infty$ performance $\gamma > 0$ is guaranteed. Furthermore, a suitable controller is of the form (10.69) with

$$\begin{aligned} \hat{A}_{ij}(\imath,\varepsilon) &= \left[Y_\varepsilon^{-1}(\imath) - X_\varepsilon(\imath)\right]^{-1}\mathcal{M}_{0_{ij}}(\imath,\varepsilon)Y_\varepsilon^{-1}(\imath) \\ \hat{B}_i(\imath) &= \left[Y_0^{-1}(\imath) - X_0(\imath)\right]^{-1}\mathcal{B}_{0_i}(\imath) \\ \hat{C}_i(\imath) &= \mathcal{C}_{0_i}(\imath)Y_0^{-1}(\imath) \end{aligned} \tag{10.80}$$

where

$$\begin{aligned} \mathcal{M}_{0_{ij}}(\imath,\varepsilon) = &-A_i^T(\imath) - X_\varepsilon(\imath)A_i(\imath)Y_\varepsilon(\imath) - \left[Y_\varepsilon^{-1}(\imath) - X_\varepsilon(\imath)\right]\hat{B}_i(\imath)C_{2_j}(\imath)Y_\varepsilon(\imath) \\ &-X_\varepsilon(\imath)B_{2_i}(\imath)\hat{C}_j(\imath)Y_\varepsilon(\imath) - \textstyle\sum_{k=1}^{s}\lambda_{\imath k}Y_\varepsilon^{-1}(k)Y_\varepsilon(\imath) \\ &-\tilde{\bar{C}}_{1_i}^T(\imath)\left[\tilde{\bar{C}}_{1_j}(\imath)Y_\varepsilon(\imath) + \tilde{\bar{D}}_{12_j}(\imath)\hat{C}_j(\imath)Y_\varepsilon(\imath)\right] \\ &-\gamma^{-2}\left\{X_\varepsilon(\imath)\tilde{\bar{B}}_{1_i}(\imath) + \left[Y_\varepsilon^{-1}(\imath) - X_\varepsilon(\imath)\right]\hat{B}_i(\imath)\tilde{\bar{D}}_{21_i}(\imath)\right\}\tilde{\bar{B}}_{1_j}^T(\imath), \end{aligned} \tag{10.81}$$

$$X_\varepsilon(\imath) = \left\{X_0(\imath) + \varepsilon\tilde{X}(\imath)\right\}E_\varepsilon \tag{10.82}$$

and

$$Y_\varepsilon^{-1}(\imath) = \left\{Y_0^{-1}(\imath) + \varepsilon N_\varepsilon(\imath)\right\}E_\varepsilon \tag{10.83}$$

with $\tilde{X}(\imath) = D\left(X_0^T(\imath) - X_0(\imath)\right)$ and $N_\varepsilon(\imath) = D\left((Y_0^{-1}(\imath))^T - Y_0^{-1}(\imath)\right)$.

Proof: Since (10.70) is of the form of (10.1), it can be shown by employing of the proof for Theorem 17. ∎

10.4 Example

Consider a modified series dc motor model based on [139] as shown in Figure 10.1 which is governed by the following differential equations:

$$\begin{aligned} J\frac{d\tilde{\omega}(t)}{dt} &= K_m L_f \tilde{i}^2(t) - (D + \Delta D)\tilde{\omega}(t) \\ L\frac{d\tilde{i}(t)}{dt} &= -R\tilde{i}(t) - K_m L_f \tilde{i}(t)\tilde{\omega}(t) + \tilde{V}(t) \end{aligned} \tag{10.84}$$

where $\tilde{\omega}(t) = \omega(t) - \omega_{ref}(t)$ is the deviation of the actual angular velocity from the desired angular velocity, $\tilde{i}(t) = i(t) - i_{ref}(t)$ is the deviation of the actual current from the desired current, $\tilde{V}(t) = V(t) - V_{ref}(t)$ is the deviation of the actual input voltage from the desired input voltage, J is the moment of inertia, K_m is the torque/back emf constant, D is the viscous friction coefficient, and R_a, R_f, L_a and L_f are the armature resistance, the field winding resistance, the armature inductance and the field winding inductance, respectively, with $R \triangleq R_f + R_a$ and $L \triangleq L_f + L_a$. Note that in a typical series-connected dc motor, the condition $L_f \gg L_a$ holds. When one obtains a series-connected dc motor, we have $i(t) = i_a(t) = i_f(t)$. Now let us assume that $|\Delta J| \leq 0.1J$.

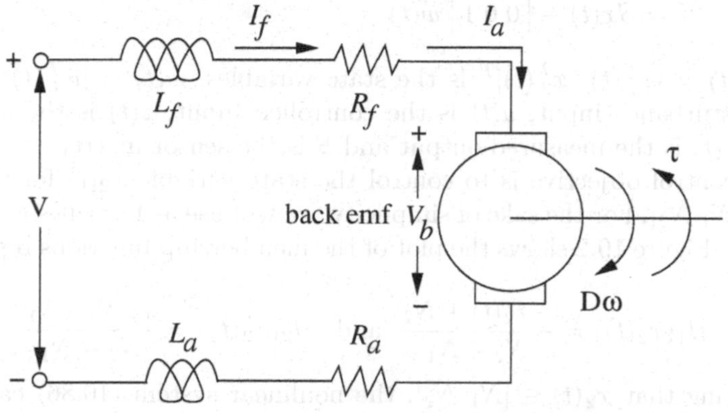

Fig. 10.1. A modified series dc motor equivalent circuit.

Giving $x_1(t) = \tilde{\omega}(t)$, $x_2(t) = \tilde{i}(t)$ and $u(t) = \tilde{V}(t)$, (10.84) becomes

$$\begin{bmatrix} \dot{x}_1(t) \\ \varepsilon\dot{x}_2(t) \end{bmatrix} = \begin{bmatrix} -\frac{D}{(J+\Delta J)} & \frac{K_m L_f}{(J+\Delta J)} x_2(t) \\ -K_m L_f x_2(t) & -R \end{bmatrix} \begin{bmatrix} x_1(t) \\ x_2(t) \end{bmatrix} + \begin{bmatrix} 0 \\ 1 \end{bmatrix} u(t) \tag{10.85}$$

where $\varepsilon = L$ represents a small parasitic parameter. Assume that, the system is aggregated into 3 modes as shown in Table 10.1 and the transition probability matrix that relates the three operation modes is given as follows:

$$P_{ik} = \begin{bmatrix} 0.67 & 0.17 & 0.16 \\ 0.30 & 0.47 & 0.23 \\ 0.26 & 0.10 & 0.64 \end{bmatrix}.$$

Table 10.1. System Terminology.

Mode $\imath$	Moment of Inertia	$J(\imath) \pm \Delta J(\imath)$ (kg·m^2)
1	Small	0.0005 ±10%
2	Normal	0.005 ±10%
3	Large	0.05 ±10%

The parameters for the system are given as $R = 10\ \Omega$, $L_f = 0.005\ H$, $D = 0.05\ N \cdot m/rad/s$ and $K_m = 1\ N \cdot m/A$. Substituting the parameters into (10.85), we get

$$\begin{aligned}
\begin{bmatrix} \dot{x}_1(t) \\ \varepsilon \dot{x}_2(t) \end{bmatrix} &= \begin{bmatrix} -\frac{0.05}{J(\imath)} & \frac{0.005}{J(\imath)} x_2(t) \\ -0.005 x_2(t) & -10 \end{bmatrix} \begin{bmatrix} x_1(t) \\ x_2(t) \end{bmatrix} + \begin{bmatrix} 0 & 0 \\ 0.1 & 0 \end{bmatrix} w(t) \\
&\quad + \begin{bmatrix} 0 \\ 1 \end{bmatrix} u(t) + \begin{bmatrix} -\frac{0.05}{\Delta J(\imath)} & \frac{0.005}{\Delta J(\imath)} x_2(t) \\ 0 & 0 \end{bmatrix} \begin{bmatrix} x_1(t) \\ x_2(t) \end{bmatrix} \\
z(t) &= \begin{bmatrix} 1 & 0 \\ 0 & 1 \end{bmatrix} \begin{bmatrix} x_1(t) \\ x_2(t) \end{bmatrix} + \begin{bmatrix} 0 \\ 1 \end{bmatrix} u(t) \\
y(t) &= Sx(t) + \begin{bmatrix} 0 & 0.1 \end{bmatrix} w(t)
\end{aligned} \tag{10.86}$$

where $x(t) = [x_1^T(t) \;\; x_2^T(t)]^T$ is the state variables, $w(t) = [w_1^T(t) \;\; w_2^T(t)]^T$ is the disturbance input, $u(t)$ is the controlled input, $z(t)$ is the controlled output, $y(t)$ is the measured output and S is the sensor matrix.

The control objective is to control the state variable $x_2(t)$ for the range $x_2(t) \in [N_1 \; N_2]$. For the sake of simplicity, we will use as few rules as possible. Note that Figure 10.2 shows the plot of the membership functions represented by

$$M_1(x_2(t)) = \frac{-x_2(t) + N_2}{N_2 - N_1} \quad \text{and} \quad M_2(x_2(t)) = \frac{x_2(t) - N_1}{N_2 - N_1}.$$

Knowing that $x_2(t) \in [N_1 \; N_2]$, the nonlinear system (10.86) can be approximated by the following TS fuzzy model:

Plant Rule 1: IF $x_2(t)$ is $M_1(x_2(t))$ THEN

$$\begin{aligned}
E_\varepsilon \dot{x}(t) &= [A_1(\imath) + \Delta A_1(\imath)]x(t) + B_{1_1}(\imath)w(t) + B_{2_1}(\imath)u(t), \quad x(0) = 0, \\
z(t) &= C_{1_1}(\imath)x(t), \\
y(t) &= C_{2_1}(\imath)x(t) + D_{21_1}(\imath)w(t).
\end{aligned}$$

Plant Rule 2: IF $x_2(t)$ is $M_2(x_2(t))$ THEN

$$\begin{aligned}
E_\varepsilon \dot{x}(t) &= [A_2(\imath) + \Delta A_2(\imath)]x(t) + B_{1_2}(\imath)w(t) + B_{2_2}(\imath)u(t), \quad x(0) = 0, \\
z(t) &= C_{1_2}(\imath)x(t), \\
y(t) &= C_{2_2}(\imath)x(t) + D_{21_2}(\imath)w(t)
\end{aligned}$$

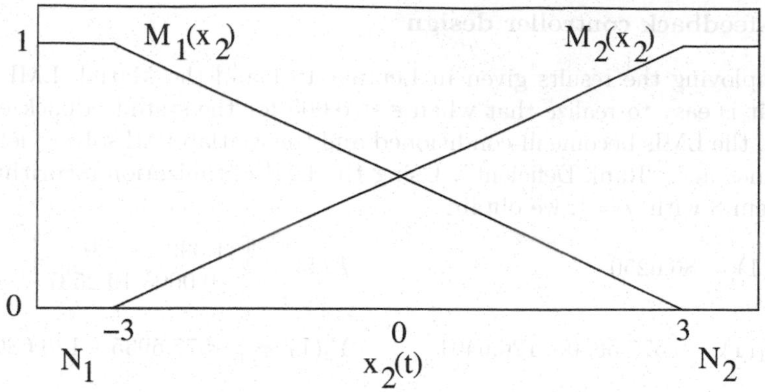

Fig. 10.2. Membership functions for the two fuzzy set.

where $x(t) = \begin{bmatrix} x_1(t) \\ x_2(t) \end{bmatrix}$, $w(t) = \begin{bmatrix} w_1(t) \\ w_2(t) \end{bmatrix}$, $E_\varepsilon = \begin{bmatrix} 1 & 0 \\ 0 & \varepsilon \end{bmatrix}$,

$$A_1(1) = \begin{bmatrix} -100 & 10N_1 \\ -0.005N_1 & -10 \end{bmatrix}, \quad A_2(1) = \begin{bmatrix} -100 & 10N_2 \\ -0.005N_2 & -10 \end{bmatrix},$$

$$A_1(2) = \begin{bmatrix} -10 & N_1 \\ -0.005N_1 & -10 \end{bmatrix}, \quad A_2(2) = \begin{bmatrix} -10 & N_2 \\ -0.005N_2 & -10 \end{bmatrix},$$

$$A_1(3) = \begin{bmatrix} -1 & 0.1N_1 \\ -0.005N_1 & -10 \end{bmatrix}, \quad A_2(3) = \begin{bmatrix} -1 & 0.1N_2 \\ -0.005N_2 & -10 \end{bmatrix},$$

$$B_{1_1}(1) = B_{1_2}(1) = B_{1_1}(2) = B_{1_2}(2) = B_{1_1}(3) = B_{1_2}(3) = \begin{bmatrix} 0 & 0 \\ 0.1 & 0 \end{bmatrix},$$

$$B_{2_1}(1) = B_{2_2}(1) = B_{2_1}(2) = B_{2_2}(2) = B_{2_1}(3) = B_{2_2}(3) = \begin{bmatrix} 0 \\ 1 \end{bmatrix},$$

$$C_{1_1}(1) = C_{1_2}(1) = C_{1_1}(2) = C_{1_2}(2) = C_{1_1}(3) = C_{1_2}(3) = \begin{bmatrix} 1 & 0 \\ 0 & 1 \end{bmatrix},$$

$$C_{2_1}(1) = C_{2_2}(1) = C_{2_1}(2) = C_{2_2}(2) = C_{2_1}(3) = C_{2_2}(3) = S,$$

$$D_{12_1}(1) = D_{12_2}(1) = D_{12_1}(2) = D_{12_2}(2) = D_{12_1}(3) = D_{12_2}(3) = \begin{bmatrix} 0 \\ 1 \end{bmatrix},$$

$$D_{21_1}(1) = D_{21_2}(1) = D_{21_1}(2) = D_{21_2}(2) = D_{21_1}(3) = D_{21_2}(3) = \begin{bmatrix} 0 & 0.1 \end{bmatrix},$$

$$\Delta A_1(\imath) = F(x(t), \imath, t)H_{1_1}(\imath) \text{ and } \Delta A_2(\imath) = F(x(t), \imath, t)H_{1_2}(\imath).$$

Now, by assuming that $\|F(x(t), \imath, t)\| \leq \rho(\imath) = 1$, we have

$$H_{1_1}(\imath) = \begin{bmatrix} -\frac{0.05}{J(\imath)} & \frac{0.05}{J(\imath)}N_1 \\ 0 & 0 \end{bmatrix} \text{ and } H_{1_2}(\imath) = \begin{bmatrix} -\frac{0.05}{J(\imath)} & \frac{0.05}{J(\imath)}N_2 \\ 0 & 0 \end{bmatrix}.$$

In this simulation, we select $N_1 = -3$ and $N_2 = 3$.

State-feedback controller design

Employing the results given in Lemma 10.1 and the Matlab LMI solver [138], it is easy to realize that when $\varepsilon < 0.005$ for the state-feedback control design, the LMIs become ill-conditioned and the Matlab LMI solver yields the error message, "Rank Deficient". Using the LMI optimization algorithm and Theorem 8 with $\gamma = 1$, we obtain

$$\delta(1) = 86.6250, \qquad P(1) = \begin{bmatrix} 1.3925 & 0 \\ -0.0003 & 19.2537 \end{bmatrix},$$

$$Y_1(1) = \begin{bmatrix} 577.5880 & -126.6401 \end{bmatrix}, \qquad Y_2(1) = \begin{bmatrix} -577.5935 & -126.6804 \end{bmatrix},$$

$$K_1(1) = \begin{bmatrix} 414.7894 & -6.5774 \end{bmatrix}, \qquad K_2(1) = \begin{bmatrix} -414.7961 & -6.5795 \end{bmatrix},$$

$$\delta(2) = 69.8795, \qquad P(2) = \begin{bmatrix} 7.5856 & 0 \\ -0.0002 & 17.5735 \end{bmatrix},$$

$$Y_1(2) = \begin{bmatrix} 52.6069 & -110.0622 \end{bmatrix}, \qquad Y_2(2) = \begin{bmatrix} -52.6069 & -110.0622 \end{bmatrix},$$

$$K_1(2) = \begin{bmatrix} 6.9351 & -6.2629 \end{bmatrix}, \qquad K_2(2) = \begin{bmatrix} -6.9351 & -6.2629 \end{bmatrix},$$

$$\delta(3) = 69.8795, \qquad P(3) = \begin{bmatrix} 3.8508 & 0 \\ -0.0001 & 17.0254 \end{bmatrix},$$

$$Y_1(3) = \begin{bmatrix} 577.5880 & -126.6401 \end{bmatrix}, \qquad Y_2(3) = \begin{bmatrix} -577.5880 & -126.6401 \end{bmatrix},$$

$$K_1(3) = \begin{bmatrix} 1.3114 & -6.1753 \end{bmatrix}, \qquad K_2(3) = \begin{bmatrix} -1.3114 & -6.1753 \end{bmatrix}.$$

The resulting fuzzy state-feedback controller is

$$u(t) = \sum_{j=1}^{2} \mu_j K_j(\imath) x(t) \tag{10.87}$$

where

$$\mu_1 = M_1(x_2(t)) \text{ and } \mu_2 = M_2(x_2(t)).$$

Output feedback controller design

Employing the results given in Lemma 10.2 and the Matlab LMI solver [138], it is easy to realize that $\varepsilon < 0.007$ for the output feedback control design in Case I and $\varepsilon < 0.008$ for the output feedback control design in Case II, the LMIs become ill-conditioned and the Matlab LMI solver yields the error message, "Rank Deficient". Using the LMI optimization algorithm and Theorems 17-18 with $\varepsilon = 0.005$, $\gamma = 1$ and $\delta(1) = \delta(2) = \delta(3) = 1$, we obtain the following results:

Case I: $\nu(t)$ are available for feedback

In this case, $x_2(t) = \nu(t)$ is assumed to be available for feedback; for instance, $S = [0 \;\; 1]$. This implies that μ_i is available for feedback.

$$X(1) = \begin{bmatrix} 8.3235 & -0.0001 \\ 0 & 3.6771 \end{bmatrix}, \qquad Y(1) = 10^4 \times \begin{bmatrix} 1.6913 & -0.0001 \\ 0 & 4.0937 \end{bmatrix},$$

$$\hat{A}_{11}(1,\varepsilon) = \begin{bmatrix} -0.0011 & -0.0036 \\ -1.6317 & -4.8951 \end{bmatrix}, \qquad \hat{A}_{12}(1,\varepsilon) = \begin{bmatrix} -0.0011 & -0.0036 \\ -1.6317 & -4.8951 \end{bmatrix},$$

$$\hat{A}_{21}(1,\varepsilon) = \begin{bmatrix} -0.0010 & -0.0036 \\ -1.6319 & -4.8952 \end{bmatrix}, \qquad \hat{A}_{22}(1,\varepsilon) = \begin{bmatrix} -0.0011 & -0.0036 \\ -1.6319 & -4.8952 \end{bmatrix},$$

$$\hat{B}_1(1) = \begin{bmatrix} -0.2703 \\ 3.0312 \end{bmatrix}, \qquad \hat{B}_2(1) = \begin{bmatrix} 0.2703 \\ 3.0312 \end{bmatrix},$$

$$\hat{C}_1(1) = \begin{bmatrix} 0.0185 & -1.7016 \end{bmatrix}, \qquad \hat{C}_2(1) = \begin{bmatrix} -0.0948 & -0.6823 \end{bmatrix},$$

$$X(2) = \begin{bmatrix} 3.6796 & -0.0001 \\ 0 & 4.1135 \end{bmatrix}, \qquad Y(2) = 10^5 \times \begin{bmatrix} 1.2364 & -0.0001 \\ 0 & 0.7850 \end{bmatrix},$$

$$\hat{A}_{11}(2,\varepsilon) = \begin{bmatrix} -0.0002 & -0.0008 \\ -0.1459 & -0.4384 \end{bmatrix}, \qquad \hat{A}_{12}(2,\varepsilon) = \begin{bmatrix} -0.0002 & -0.0008 \\ -0.1459 & -0.4384 \end{bmatrix},$$

$$\hat{A}_{21}(2,\varepsilon) = \begin{bmatrix} -0.0002 & -0.0009 \\ -0.1460 & -0.4387 \end{bmatrix}, \qquad \hat{A}_{22}(2,\varepsilon) = \begin{bmatrix} -0.0002 & -0.0009 \\ -0.1460 & -0.4387 \end{bmatrix},$$

$$\hat{B}_1(2) = \begin{bmatrix} -0.0448 \\ 0.1124 \end{bmatrix}, \qquad \hat{B}_2(2) = \begin{bmatrix} 0.0448 \\ 0.1124 \end{bmatrix},$$

$$\hat{C}_1(2) = \begin{bmatrix} 0.0039 & -0.0839 \end{bmatrix}, \qquad \hat{C}_2(2) = \begin{bmatrix} -0.0030 & -0.0972 \end{bmatrix},$$

$$X(3) = \begin{bmatrix} 0.2416 & -0.0001 \\ 0 & 4.1502 \end{bmatrix}, \qquad Y(3) = 10^5 \times \begin{bmatrix} 9.9738 & 0.9206 \\ 0 & 1.0041 \end{bmatrix},$$

$$\hat{A}_{11}(3,\varepsilon) = \begin{bmatrix} -0.0011 & -0.0012 \\ -0.0145 & -0.0517 \end{bmatrix}, \qquad \hat{A}_{12}(3,\varepsilon) = \begin{bmatrix} -0.0011 & -0.0012 \\ -0.0145 & -0.0517 \end{bmatrix},$$

$$\hat{A}_{21}(3,\varepsilon) = \begin{bmatrix} -0.0011 & -0.0014 \\ -0.0148 & -0.0518 \end{bmatrix}, \qquad \hat{A}_{22}(3,\varepsilon) = \begin{bmatrix} -0.0011 & -0.0014 \\ -0.0148 & -0.0518 \end{bmatrix},$$

$$\hat{B}_1(3) = \begin{bmatrix} -0.0070 \\ 0.0098 \end{bmatrix}, \qquad \hat{B}_2(3) = \begin{bmatrix} 0.0070 \\ 0.0098 \end{bmatrix},$$

$$\hat{C}_1(3) = \begin{bmatrix} 0.0036 & -0.0266 \end{bmatrix}, \qquad \hat{C}_2(3) = \begin{bmatrix} 0.0036 & -0.0244 \end{bmatrix}.$$

The resulting fuzzy controller is

$$\begin{array}{ll} E_\varepsilon \dot{\hat{x}}(t) & = \sum_{i=1}^{2}\sum_{j=1}^{2}\mu_i\mu_j\hat{A}_{ij}(\imath,\varepsilon)\hat{x}(t) + \sum_{i=1}^{2}\mu_i\hat{B}_i(\imath)y(t) \\ u(t) & = \sum_{i=1}^{2}\mu_i\hat{C}_i(\imath)\hat{x}(t) \end{array} \tag{10.88}$$

where

$$\mu_1 = M_1(x_2(t)) \text{ and } \mu_2 = M_2(x_2(t)).$$

Case II: $\nu(t)$ are unavailable for feedback

In this case, $x_2(t) = \nu(t)$ is assumed to be unavailable for feedback; for instance, $S = [1 \;\; 0]$. This implies that μ_i is unavailable for feedback.

$$X(1) = \begin{bmatrix} 5.1464 & 0.0015 \\ 0 & 2.8529 \end{bmatrix}, \qquad Y(1) = 10^4 \times \begin{bmatrix} 1.9180 & -0.0001 \\ 0 & 4.6424 \end{bmatrix},$$

$$\hat{A}_{11}(1,\varepsilon) = \begin{bmatrix} -0.0018 & -0.0058 \\ -2.1031 & -6.3093 \end{bmatrix}, \qquad \hat{A}_{12}(1,\varepsilon) = \begin{bmatrix} -0.0018 & -0.0058 \\ -2.1031 & -6.3093 \end{bmatrix},$$

$$\hat{A}_{21}(1,\varepsilon) = \begin{bmatrix} -0.0019 & -0.0058 \\ -2.1035 & -6.3094 \end{bmatrix}, \qquad \hat{A}_{22}(1,\varepsilon) = \begin{bmatrix} -0.0019 & -0.0058 \\ -2.1035 & -6.3094 \end{bmatrix},$$

$$\hat{B}_1(1) = \begin{bmatrix} 0.1238 \\ -0.7738 \end{bmatrix}, \qquad \hat{B}_2(1) = \begin{bmatrix} 0.1234 \\ 0.7914 \end{bmatrix},$$

$$\hat{C}_1(1) = \begin{bmatrix} 0.0473 & -1.4904 \end{bmatrix}, \qquad \hat{C}_2(1) = \begin{bmatrix} -0.1964 & -0.5662 \end{bmatrix},$$

$$X(2) = \begin{bmatrix} 2.9941 & 0.0004 \\ 0 & 3.3210 \end{bmatrix}, \qquad Y(2) = 10^5 \times \begin{bmatrix} 1.4022 & -0.0001 \\ 0 & 0.8902 \end{bmatrix},$$

$$\hat{A}_{11}(2,\varepsilon) = \begin{bmatrix} -0.0002 & -0.0010 \\ -0.1807 & -0.5430 \end{bmatrix}, \qquad \hat{A}_{12}(2,\varepsilon) = \begin{bmatrix} -0.0002 & -0.0010 \\ -0.1807 & -0.5430 \end{bmatrix},$$

$$\hat{A}_{21}(2,\varepsilon) = \begin{bmatrix} -0.0002 & -0.0012 \\ -0.1808 & -0.5431 \end{bmatrix}, \qquad \hat{A}_{22}(2,\varepsilon) = \begin{bmatrix} -0.0002 & -0.0012 \\ -0.1808 & -0.5431 \end{bmatrix},$$

$$\hat{B}_1(2) = \begin{bmatrix} 0.1359 \\ -0.0372 \end{bmatrix}, \qquad \hat{B}_2(2) = \begin{bmatrix} 0.1359 \\ 0.0461 \end{bmatrix},$$

$$\hat{C}_1(2) = \begin{bmatrix} 0.0028 & -0.0815 \end{bmatrix}, \qquad \hat{C}_2(2) = \begin{bmatrix} -0.0021 & -0.0912 \end{bmatrix},$$

$$X(3) = \begin{bmatrix} 0.2471 & 0.0001 \\ 0 & 3.3689 \end{bmatrix}, \qquad Y(3) = 10^6 \times \begin{bmatrix} 1.1311 & 0.1044 \\ 0 & 0.1139 \end{bmatrix},$$

$$\hat{A}_{11}(3,\varepsilon) = \begin{bmatrix} -0.0005 & -0.0012 \\ -0.0178 & -0.0639 \end{bmatrix}, \qquad \hat{A}_{12}(3,\varepsilon) = \begin{bmatrix} -0.0005 & -0.0012 \\ -0.0178 & -0.0639 \end{bmatrix},$$

$$\hat{A}_{21}(3,\varepsilon) = \begin{bmatrix} -0.0006 & -0.0011 \\ -0.0178 & -0.0641 \end{bmatrix}, \qquad \hat{A}_{22}(3,\varepsilon) = \begin{bmatrix} -0.0006 & -0.0011 \\ -0.0178 & -0.0641 \end{bmatrix},$$

$$\hat{B}_1(3) = \begin{bmatrix} 0.1591 \\ -0.0025 \end{bmatrix}, \qquad \hat{B}_2(3) = \begin{bmatrix} 0.1591 \\ 0.0047 \end{bmatrix},$$

$$\hat{C}_1(3) = \begin{bmatrix} 0.0025 & -0.0363 \end{bmatrix}, \qquad \hat{C}_2(3) = \begin{bmatrix} 0.0025 & -0.0347 \end{bmatrix}.$$

The resulting fuzzy controller is

$$\begin{aligned} E_\varepsilon \dot{\hat{x}}(t) &= \textstyle\sum_{i=1}^{2} \sum_{j=1}^{2} \hat{\mu}_i \hat{\mu}_j \hat{A}_{ij}(\imath,\varepsilon)\hat{x}(t) + \sum_{i=1}^{2} \hat{\mu}_i \hat{B}_i(\imath) y(t) \\ u(t) &= \textstyle\sum_{i=1}^{2} \hat{\mu}_i \hat{C}_i(\imath)\hat{x}(t) \end{aligned} \tag{10.89}$$

where

$$\hat{\mu}_1 = M_1(\hat{x}_2(t)) \text{ and } \hat{\mu}_2 = M_2(\hat{x}_2(t)).$$

Remark 12. For a sufficiently small ε, both robust fuzzy state and output feedback controllers guarantee that the $\mathcal{L}_2$-gain, γ, is less than the prescribed value. Figure 10.3 shows the result of the changing between modes during the simulation with the initial mode 1 and $\varepsilon = 0.005$. The disturbance input signal, $w(t)$, which was used during the simulation is given in Figure 10.4. The ratio of the regulated output energy to the disturbance input noise energy obtained by using the $\mathcal{H}_\infty$ fuzzy controllers is depicted in Figure 10.5. The ratio of the regulated output energy to the disturbance input noise energy tends to a constant value which is about 0.0097 *for the state-feedback controller, and* 0.0157 *for the output feedback controller in Case I and* 0.0162 *for the output feedback controller in Case II . So $\gamma = \sqrt{0.0097} = 0.0985$ for the state-feedback controller, and $\gamma = \sqrt{0.0157} = 0.1259$ for the output feedback controller in Case I and $\gamma = \sqrt{0.0162} = 0.1273$ for the output feedback controller in Case II which all are less than the prescribed value* 1. *Finally, Table 10.2 shows the performance index, γ, for different values of ε.* □

Table 10.2. The performance index γ of the system with different values of ε.

	The performance index γ		
ε	State-feedback	Output-feedback in Case I	Output-feedback in Case II
0.005	0.0985	0.1259	0.1273
0.10	0.4796	0.5657	0.5831
0.31	0.8660	0.9643	0.9945
0.32	0.8832	0.9899	>1
0.33	0.8944	>1	>1
0.40	0.9945	>1	>1
0.41	>1	>1	>1

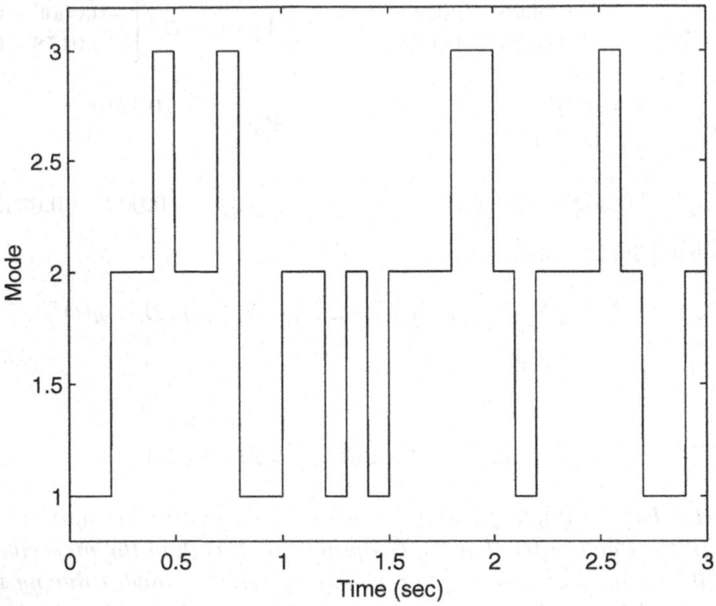

Fig. 10.3. The result of the changing between modes during the simulation with the initial mode 1.

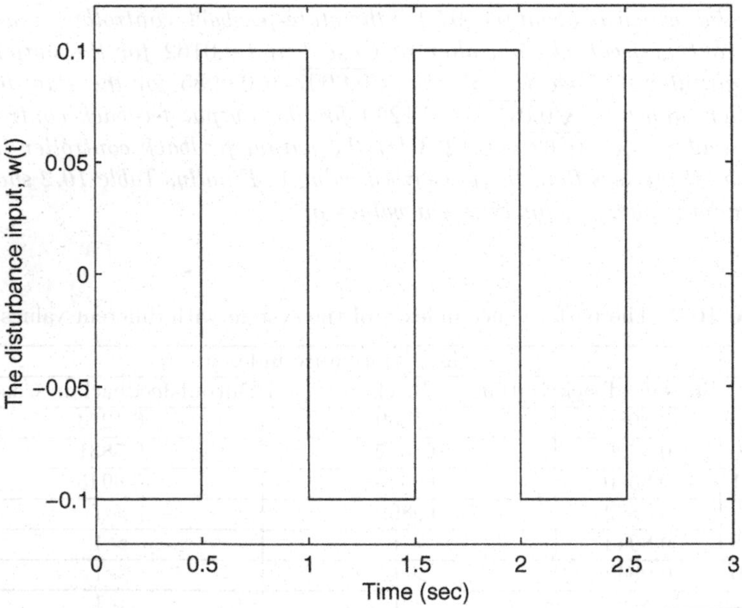

Fig. 10.4. The disturbance input, $w(t)$.

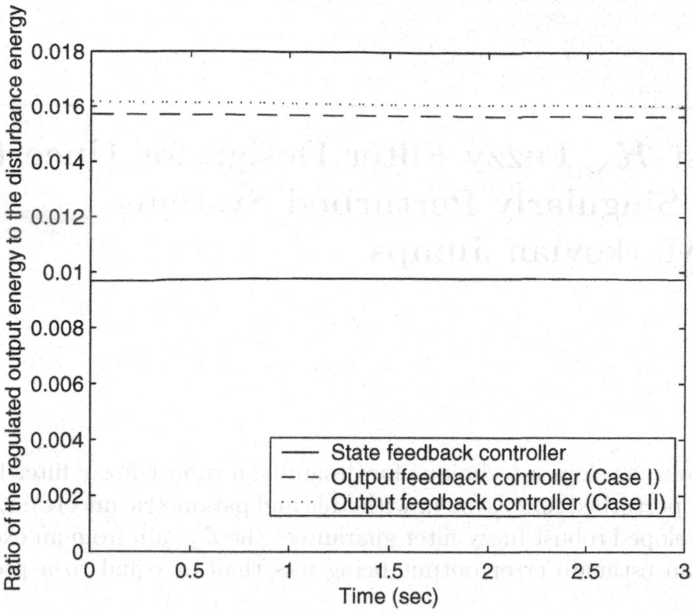

Fig. 10.5. The ratio of the regulated output energy to the disturbance noise energy, $\left(\frac{\int_0^{T_f} z^T(t)z(t)dt}{\int_0^{T_f} w^T(t)w(t)dt}\right)$.

10.5 Conclusion

A complete methodology of designing a robust fuzzy state and output feedback controller for a TS singularly perturbed fuzzy system with MJs and parametric uncertainties has been proposed. The design approach does not involve the separation of states into slow and fast subsystems and it can be applied not only to standard, but also to nonstandard UFSPS–MJs. Sufficient conditions for the existence of the robust $\mathcal{H}_\infty$ fuzzy controllers have been derived in terms of a family of ε-independent LMIs.

11

Robust $\mathcal{H}_\infty$ Fuzzy Filter Design for Uncertain Fuzzy Singularly Perturbed Systems with Markovian Jumps

This chapter develops a technique for designing a robust fuzzy filter for a TS singularly perturbed fuzzy system with MJs and parametric uncertainties. The newly developed robust fuzzy filter guarantees the $\mathcal{L}_2$-gain from an exogenous input to an estimate error output being less than or equal to a prescribed value.

11.1 Robust $\mathcal{H}_\infty$ Fuzzy Filter Design

In this chapter, without loss of generality, we assume $u(t) = 0$. Let us recall the system (10.1) with $u(t) = 0$ as follows:

$$\begin{aligned} E_\varepsilon \dot{x}(t) &= \textstyle\sum_{i=1}^r \mu_i \Big[[A_i(\eta) + \Delta A_i(\eta)]x(t) + [B_{1_i}(\eta) + \Delta B_{1_i}(\eta)]w(t)\Big], x(0) = 0, \\ z(t) &= \textstyle\sum_{i=1}^r \mu_i \Big[[C_{1_i}(\eta) + \Delta C_{1_i}(\eta)]x(t)\Big] \\ y(t) &= \textstyle\sum_{i=1}^r \mu_i \Big[[C_{2_i}(\eta) + \Delta C_{2_i}(\eta)]x(t) + [D_{21_i}(\eta) + \Delta D_{21_i}(\eta)]w(t)\Big]. \end{aligned} \tag{11.1}$$

The aim is to design a full order dynamic $\mathcal{H}_\infty$ fuzzy filter of the form

$$\begin{aligned} E_\varepsilon \dot{\hat{x}}(t) &= \textstyle\sum_{i=1}^r \sum_{j=1}^r \hat{\mu}_i \hat{\mu}_j \Big[\hat{A}_{ij}(\imath, \varepsilon)\hat{x}(t) + \hat{B}_i(\imath)y(t)\Big] \\ \hat{z}(t) &= \textstyle\sum_{i=1}^r \hat{\mu}_i \hat{C}_i(\imath)\hat{x}(t) \end{aligned} \tag{11.2}$$

where $\hat{x}(t) \in \Re^n$ is the filter's state vector, $\hat{z} \in \Re^s$ is the estimate of $z(t)$, $\hat{A}_{ij}(\imath, \varepsilon)$, $\hat{B}_i(\imath)$ and $\hat{C}_i(\imath)$ are parameters of the filter which are to be determined, and $\hat{\mu}_i$ denotes the normalized time-varying fuzzy weighting functions for each rule (i.e., $\hat{\mu}_i \geq 0$ and $\sum_{i=1}^r \hat{\mu}_i = 1$), such that the inequality (6.3) holds. Clearly, in real control problems, all of the premise variables are not necessarily measurable. Thus, in this section, we consider the designing of the robust $\mathcal{H}_\infty$ fuzzy filter into two cases as follows. Subsection 11.1.1 considers the case where the premise variable of the fuzzy model μ_i is measurable, while in Subsection 11.1.2, the premise variable is assumed to be unmeasurable.

11.1.1 Case I–$\nu(t)$ is available for feedback

The premise variable of the fuzzy model $\nu(t)$ is available for feedback which implies that μ_i is available for feedback. Thus, we can select our filter that depends on μ_i as follows [91]:

$$\begin{aligned} E_\varepsilon \dot{\hat{x}}(t) &= \sum_{i=1}^r \sum_{j=1}^r \mu_i \mu_j \left[\hat{A}_{ij}(\imath,\varepsilon)\hat{x}(t) + \hat{B}_i(\imath)y(t) \right] \\ \hat{z}(t) &= \sum_{i=1}^r \mu_i \hat{C}_i(\imath)\hat{x}(t). \end{aligned} \tag{11.3}$$

Before presenting our next result, the following lemma is recalled.

Lemma 10. *Consider the system (11.1). Given a prescribed $\mathcal{H}_\infty$ performance $\gamma > 0$ and any positive constants $\delta(\imath)$, for $\imath = 1, 2, \cdots, s$, if there exist matrices $X_\varepsilon(\imath) = X_\varepsilon^T(\imath)$, $Y_\varepsilon(\imath) = Y_\varepsilon^T(\imath)$, $\mathcal{B}_i(\imath,\varepsilon)$ and $\mathcal{C}_i(\imath,\varepsilon)$, $i = 1, 2, \cdots, r$, satisfying the following ε-dependent linear matrix inequalities:*

$$\begin{bmatrix} X_\varepsilon(\imath) & I \\ I & Y_\varepsilon(\imath) \end{bmatrix} > 0 \tag{11.4}$$

$$X_\varepsilon(\imath) > 0 \tag{11.5}$$

$$Y_\varepsilon(\imath) > 0 \tag{11.6}$$

$$\Psi_{11_{ii}}(\imath,\varepsilon) < 0, \quad i = 1, 2, \cdots, r \tag{11.7}$$

$$\Psi_{22_{ii}}(\imath,\varepsilon) < 0, \quad i = 1, 2, \cdots, r \tag{11.8}$$

$$\Psi_{11_{ij}}(\imath,\varepsilon) + \Psi_{11_{ji}}(\imath,\varepsilon) < 0, \quad i < j \le r \tag{11.9}$$

$$\Psi_{22_{ij}}(\imath,\varepsilon) + \Psi_{22_{ji}}(\imath,\varepsilon) < 0, \quad i < j \le r \tag{11.10}$$

where

$$\Psi_{11_{ij}}(\imath) = \begin{pmatrix} \begin{pmatrix} E_\varepsilon^{-1}A_i(\imath)Y_\varepsilon(\imath) + Y_\varepsilon(\imath)A_i^T(\imath)E_\varepsilon^{-1} \\ +\gamma^{-2}E_\varepsilon^{-1}\tilde{B}_{1_i}(\imath)\tilde{B}_{1_j}^T(\imath)E_\varepsilon^{-1} \\ +\lambda_{\imath\imath}Y_\varepsilon(\imath)E_\varepsilon^{-1} \end{pmatrix} & (*)^T & (*)^T \\ \tilde{C}_{1_i}(\imath)Y_\varepsilon(\imath) + E_\varepsilon^{-1}\tilde{D}_{12}(\imath)\mathcal{C}_j(\imath,\varepsilon) & -I & (*)^T \\ \mathcal{J}^T(\imath) & 0 & -\mathcal{Y}_\varepsilon(\imath) \end{pmatrix} \tag{11.11}$$

$$\Psi_{22_{ij}}(\imath,\varepsilon) = \begin{pmatrix} \begin{pmatrix} A_i^T(\imath)E_\varepsilon^{-1}X_\varepsilon(\imath) + X_\varepsilon(\imath)E_\varepsilon^{-1}A_i(\imath) \\ +\mathcal{B}_i(\imath,\varepsilon)C_{2_j}(\imath) + C_{2_i}^T(\imath)\mathcal{B}_j^T(\imath,\varepsilon) \\ +\tilde{C}_{1_i}^T(\imath)\tilde{C}_{1_j}(\imath) + \sum_{k=1}^s \lambda_{\imath k}X_\varepsilon(k)E_\varepsilon^{-1} \end{pmatrix} & (*)^T \\ \left[X_\varepsilon(\imath)E_\varepsilon^{-1}\tilde{B}_{1_i}(\imath) + \mathcal{B}_i(\imath,\varepsilon)\tilde{D}_{21_j}(\imath)\right]^T & -\gamma^2 I \end{pmatrix} \tag{11.12}$$

with

$$\mathcal{J}(\imath) = \left[\sqrt{\lambda_{1\imath}}Y_\varepsilon(\imath) \ \cdots \ \sqrt{\lambda_{(i-1)\imath}}Y_\varepsilon(\imath) \ \sqrt{\lambda_{(i+1)\imath}}Y_\varepsilon(\imath) \ \cdots \ \sqrt{\lambda_{s\imath}}Y_\varepsilon(\imath)\right]$$

$$\mathcal{Y}_\varepsilon(\imath) = diag\left\{Y_\varepsilon(1), \ \cdots, \ Y_\varepsilon(\imath-1), \ Y_\varepsilon(\imath+1), \ \cdots \ Y_\varepsilon(s)\right\}$$

$$\tilde{B}_{1_i}(\imath) = [\delta(\imath)I \ \ I \ \ 0 \ \ B_{1_i}(\imath) \ \ 0]$$

$$\tilde{C}_{1_i}(\imath) = \left[\tfrac{\gamma\rho(\imath)}{\delta(\imath)}H_{1_i}^T(\imath) \ \ \tfrac{\gamma\rho(\imath)}{\delta(\imath)}H_{5_i}^T(\imath) \ \ \sqrt{2}\aleph(\imath)\rho(\imath)H_{4_i}^T(\imath) \ \ \sqrt{2}\aleph(\imath)C_{1_i}^T(\imath)\right]^T$$

$$\tilde{D}_{12}(\imath) = \begin{bmatrix} 0 & 0 & 0 & -\sqrt{2}\aleph(\imath)I \end{bmatrix}^T$$
$$\tilde{D}_{21_i}(\imath) = \begin{bmatrix} 0 & 0 & \delta(\imath)I & D_{21_i}(\imath) & I \end{bmatrix}$$
$$\aleph(\imath) = \left(1 + \rho^2(\imath)\sum_{i=1}^{r}\sum_{j=1}^{r}\left[\|H_{2_i}^T(\imath)H_{2_j}(\imath)\| + \|H_{7_i}^T(\imath)H_{7_j}(\imath)\|\right]\right)^{\frac{1}{2}},$$

then the prescribed $\mathcal{H}_\infty$ performance $\gamma > 0$ is guaranteed. Furthermore, a suitable filter is of the form (11.3) with

$$\begin{aligned} \hat{A}_{ij}(\imath,\varepsilon) &= E_\varepsilon\left[Y_\varepsilon^{-1}(\imath) - X_\varepsilon(\imath)\right]^{-1}\mathcal{M}_{ij}(\imath,\varepsilon)Y_\varepsilon^{-1}(\imath) \\ \hat{B}_i(\imath) &= E_\varepsilon\left[Y_\varepsilon^{-1}(\imath) - X_\varepsilon(\imath)\right]^{-1}\mathcal{B}_i(\imath,\varepsilon) \\ \hat{C}_i(\imath) &= \mathcal{C}_i(\imath,\varepsilon)E_\varepsilon^{-1}Y_\varepsilon^{-1}(\imath) \end{aligned} \tag{11.13}$$

where

$$\begin{aligned} &\mathcal{M}_{ij}(\imath,\varepsilon) = -A_i^T(\imath)E_\varepsilon^{-1} - X_\varepsilon(\imath)E_\varepsilon^{-1}A_i(\imath)Y_\varepsilon(\imath) \\ &-\left[Y_\varepsilon^{-1}(\imath) - X_\varepsilon(\imath)\right]E_\varepsilon^{-1}\hat{B}_i(\imath)C_{2_j}(\imath)Y_\varepsilon(\imath) \\ &-\textstyle\sum_{k=1}^{s}\lambda_{\imath k}Y_\varepsilon^{-1}(k)Y_\varepsilon(\imath) - \tilde{C}_{1_i}^T(\imath)\left[\tilde{C}_{1_j}(\imath)Y_\varepsilon(\imath) + \tilde{D}_{12_j}(\imath)\hat{C}_j(\imath)Y_\varepsilon(\imath)\right] \\ &-\gamma^{-2}\left\{X_\varepsilon(\imath)E_\varepsilon^{-1}\tilde{B}_{1_i}(\imath) + \left[Y_\varepsilon^{-1}(\imath) - X_\varepsilon(\imath)\right]E_\varepsilon^{-1}\hat{B}_i(\imath)\tilde{D}_{21_i}(\imath)\right\}\tilde{B}_{1_j}^T(\imath)E_\varepsilon^{-1}. \end{aligned} \tag{11.14}$$

Proof: The proof can be carried out by the same technique used in Lemma 6.1 and Theorem 6.1. ■

Remark 13. The LMIs given in Lemma 10 may become ill-conditioned when ε is sufficiently small, which is always the case for the SPS–MJ. In general, these ill-conditioned LMIs are very difficult to solve. Thus, to alleviate these ill-conditioned LMIs, we have the following ε-independent well-posed LMI-based sufficient conditions for the UFSPS–MJs to obtain the prescribed $\mathcal{H}_\infty$ performance. □

Theorem 19. *Consider the system (11.1). Given a prescribed $\mathcal{H}_\infty$ performance $\gamma > 0$ and any positive constants $\delta(\imath)$, for $\imath = 1, 2, \cdots, s$, if there exist matrices $X_0(\imath)$, $Y_0(\imath)$, $\mathcal{B}_{0_i}(\imath)$ and $\mathcal{C}_{0_i}(\imath)$, $i = 1, 2, \cdots, r$, satisfying the following ε-independent linear matrix inequalities:*

$$\begin{bmatrix} X_0(\imath)E + DX_0(\imath) & I \\ I & Y_0(\imath)E + DY_0(\imath) \end{bmatrix} > 0 \quad (11.15)$$

$$EX_0^T(\imath) = X_0(\imath)E, \;\; X_0^T(\imath)D = DX_0(\imath), \;\; X_0(\imath)E + DX_0(\imath) > 0 \quad (11.16)$$

$$EY_0^T(\imath) = Y_0(\imath)E, \;\; Y_0^T(\imath)D = DY_0(\imath), \;\; Y_0(\imath)E + DY_0(\imath) > 0 \quad (11.17)$$

$$\Psi_{11_{ii}}(\imath) < 0, \quad i = 1, 2, \cdots, r \quad (11.18)$$

$$\Psi_{22_{ii}}(\imath) < 0, \quad i = 1, 2, \cdots, r \quad (11.19)$$

$$\Psi_{11_{ij}}(\imath) + \Psi_{11_{ji}}(\imath) < 0, \quad i < j \le r \quad (11.20)$$

$$\Psi_{22_{ij}}(\imath) + \Psi_{22_{ji}}(\imath) < 0, \quad i < j \le r \quad (11.21)$$

where $E = \begin{pmatrix} I & 0 \\ 0 & 0 \end{pmatrix}$, $D = \begin{pmatrix} 0 & 0 \\ 0 & I \end{pmatrix}$,

$$\Psi_{11_{ij}}(\imath) = \begin{pmatrix} \begin{pmatrix} A_i(\imath)Y_0^T(\imath) + Y_0(\imath)A_i^T(\imath) \\ +\lambda_{\imath\imath}\bar{\bar{Y}}_0(\imath) + \gamma^{-2}\tilde{B}_{1_i}(\imath)\tilde{B}_{1_j}^T(\imath) \end{pmatrix} & (*)^T & (*)^T \\ \tilde{C}_{1_i}(\imath)Y_0^T(\imath) + \tilde{D}_{12}(\imath)\mathcal{C}_{0_j}(\imath) & -I & (*)^T \\ \mathcal{J}_0^T(\imath) & 0 & -\mathcal{Y}_0(\imath) \end{pmatrix} \quad (11.22)$$

$$\Psi_{22_{ij}}(\imath) = \begin{pmatrix} \begin{pmatrix} A_i^T(\imath)X_0^T(\imath) + X_0(\imath)A_i(\imath) \\ +\mathcal{B}_{0_i}(\imath)C_{2_j}(\imath) + C_{2_i}^T(\imath)\mathcal{B}_{0_j}^T(\imath) \\ +\tilde{C}_{1_i}^T(\imath)\tilde{C}_{1_j}(\imath) + \sum_{k=1}^{s}\lambda_{\imath k}\bar{\bar{X}}_0(k) \end{pmatrix} & (*)^T \\ \tilde{B}_{1_i}^T(\imath)X_0^T(\imath) + \tilde{D}_{21_i}^T(\imath)\mathcal{B}_{0_j}^T(\imath) & -\gamma^2 I \end{pmatrix} \quad (11.23)$$

with $\bar{\bar{X}}_0(k) = \frac{X_0(k)+X_0^T(k)}{2}$, $\bar{\bar{Y}}_0(\imath) = \frac{Y_0(\imath)+Y_0^T(\imath)}{2}$,

$$\mathcal{J}_0(\imath) = \left[\sqrt{\lambda_{1\imath}}\bar{\bar{Y}}_0(\imath) \cdots \sqrt{\lambda_{(i-1)\imath}}\bar{\bar{Y}}_0(\imath) \; \sqrt{\lambda_{(i+1)\imath}}\bar{\bar{Y}}_0(\imath) \cdots \sqrt{\lambda_{s\imath}}\bar{\bar{Y}}(\imath)\right]$$

$$\mathcal{Y}_0(\imath) = diag\left\{\bar{\bar{Y}}_0(1), \; \cdots, \; \bar{\bar{Y}}_0(\imath-1), \; \bar{\bar{Y}}_0(\imath+1), \; \cdots, \bar{\bar{Y}}_0(s)\right\}$$

$$\tilde{B}_{1_i}(\imath) = [\delta(\imath)I \;\; I \;\; 0 \;\; B_{1_i}(\imath) \;\; 0]$$

$$\tilde{C}_{1_i}(\imath) = \left[\frac{\gamma\rho(\imath)}{\delta(\imath)}H_{1_i}^T(\imath) \;\; \frac{\gamma\rho(\imath)}{\delta(\imath)}H_{5_i}^T(\imath) \;\; \sqrt{2}\aleph(\imath)\rho(\imath)H_{4_i}^T(\imath) \;\; \sqrt{2}\aleph(\imath)C_{1_i}^T(\imath)\right]^T$$

$$\tilde{D}_{12}(\imath) = \left[0 \;\; 0 \;\; 0 \;\; -\sqrt{2}\aleph(\imath)I\right]^T$$

$$\tilde{D}_{21_i}(\imath) = [0 \;\; 0 \;\; \delta(\imath)I \;\; D_{21_i}(\imath) \;\; I]$$

$$\aleph(\imath) = \left(1 + \rho^2(\imath)\sum_{i=1}^{r}\sum_{j=1}^{r}\left[\|H_{2_i}^T(\imath)H_{2_j}(\imath)\| + \|H_{7_i}^T(\imath)H_{7_j}(\imath)\|\right]\right)^{\frac{1}{2}},$$

then there exists a sufficiently small $\hat{\varepsilon} > 0$ *such that for* $\varepsilon \in (0, \hat{\varepsilon}]$, *the prescribed* $\mathcal{H}_\infty$ *performance* $\gamma > 0$ *is guaranteed. Furthermore, a suitable filter is of the form (11.3) with*

$$\begin{aligned} \hat{A}_{ij}(\imath,\varepsilon) &= \left[Y_\varepsilon^{-1}(\imath) - X_\varepsilon(\imath)\right]^{-1}\mathcal{M}_{0_{ij}}(\imath,\varepsilon)Y_\varepsilon^{-1}(\imath) \\ \hat{B}_i(\imath) &= \left[Y_0^{-1}(\imath) - X_0(\imath)\right]^{-1}\mathcal{B}_{0_i}(\imath) \\ \hat{C}_i(\imath) &= \mathcal{C}_{0_i}(\imath)Y_0^{-1}(\imath) \end{aligned} \quad (11.24)$$

where

$$\begin{aligned}\mathcal{M}_{0_{ij}}(\imath,\varepsilon) = & -A_i^T(\imath) - X_\varepsilon(\imath)A_i(\imath)Y_\varepsilon(\imath) - \left[Y_\varepsilon^{-1}(\imath) - X_\varepsilon(\imath)\right]\hat{B}_i(\imath)C_{2_j}(\imath)Y_\varepsilon(\imath) \\ & - \textstyle\sum_{k=1}^{s} \lambda_{\imath k} Y_\varepsilon^{-1}(k)Y_\varepsilon(\imath) - \tilde{C}_{1_i}^T(\imath)\left[\tilde{C}_{1_j}(\imath)Y_\varepsilon(\imath) + \tilde{D}_{12}(\imath)\hat{C}_j(\imath)Y_\varepsilon(\imath)\right] \\ & -\gamma^{-2}\left\{X_\varepsilon(\imath)\tilde{B}_{1_i}(\imath) + \left[Y_\varepsilon^{-1}(\imath) - X_\varepsilon(\imath)\right]\hat{B}_i(\imath)\tilde{D}_{21_i}(\imath)\right\}\tilde{B}_{1_j}^T(\imath),\end{aligned} \tag{11.25}$$

$$X_\varepsilon(\imath) = \left\{X_0(\imath) + \varepsilon\tilde{X}(\imath)\right\}E_\varepsilon \tag{11.26}$$

and

$$Y_\varepsilon^{-1}(\imath) = \left\{Y_0^{-1}(\imath) + \varepsilon N_\varepsilon(\imath)\right\}E_\varepsilon \tag{11.27}$$

with $\tilde{X}(\imath) = D\Big(X_0^T(\imath) - X_0(\imath)\Big)$ *and* $N_\varepsilon(\imath) = D\Big((Y_0^{-1}(\imath))^T - Y_0^{-1}(\imath)\Big)$.

Proof: The proof can be carried out by a similar technique used in Theorem 17. ∎

11.1.2 Case II–$\nu(t)$ is unavailable for feedback

Now, the premise variable of the fuzzy model $\nu(t)$ is unavailable for feedback which implies μ_i is unavailable for feedback. Hence, we cannot select our filter which depends on μ_i. Thus, we select our filter as follows [91]:

$$\begin{aligned} E_\varepsilon\dot{\hat{x}}(t) &= \textstyle\sum_{i=1}^{r}\sum_{j=1}^{r}\hat{\mu}_i\hat{\mu}_j\left[\hat{A}_{ij}(\imath,\varepsilon)\hat{x}(t) + \hat{B}_i(\imath)y(t)\right] \\ \hat{z}(t) &= \textstyle\sum_{i=1}^{r}\hat{\mu}_i\hat{C}_i(\imath)\hat{x}(t)\end{aligned} \tag{11.28}$$

where $\hat{\mu}_i$ depends on the premise variable of the filter which is different from μ_i.

By applying the same technique used in Subsection 3.3.2, we have the following theorem.

Theorem 20. *Consider the system (11.1). Given a prescribed $\mathcal{H}_\infty$ performance $\gamma > 0$ and any positive constants $\delta(\imath)$, for $\imath = 1, 2, \cdots, s$, if there exist matrices $X_0(\imath)$, $Y_0(\imath)$, $\mathcal{B}_{0_i}(\imath)$ and $\mathcal{C}_{0_i}(\imath)$, $i = 1, 2, \cdots, r$, satisfying the following ε-independent linear matrix inequalities:*

$$\begin{bmatrix} X_0(\imath)E + DX_0(\imath) & I \\ I & Y_0(\imath)E + DY_0(\imath)\end{bmatrix} > 0 \tag{11.29}$$

$$EX_0^T(\imath) = X_0(\imath)E,\ \ X_0^T(\imath)D = DX_0(\imath),\ \ X_0(\imath)E + DX_0(\imath) > 0 \tag{11.30}$$

$$EY_0^T(\imath) = Y_0(\imath)E,\ \ Y_0^T(\imath)D = DY_0(\imath),\ \ Y_0(\imath)E + DY_0(\imath) > 0 \tag{11.31}$$

$$\Psi_{11_{ii}}(\imath) < 0, \quad i = 1, 2, \cdots, r \tag{11.32}$$

$$\Psi_{22_{ii}}(\imath) < 0, \quad i = 1, 2, \cdots, r \tag{11.33}$$

$$\Psi_{11_{ij}}(\imath) + \Psi_{11_{ji}}(\imath) < 0, \quad i < j \le r \tag{11.34}$$

$$\Psi_{22_{ij}}(\imath) + \Psi_{22_{ji}}(\imath) < 0, \quad i < j \le r \tag{11.35}$$

where $E = \begin{pmatrix} I & 0 \\ 0 & 0 \end{pmatrix}$, $D = \begin{pmatrix} 0 & 0 \\ 0 & I \end{pmatrix}$,

$$\Psi_{11_{ij}}(\imath) = \begin{pmatrix} \begin{pmatrix} A_i(\imath)Y_0^T(\imath) + Y_0(\imath)A_i^T(\imath) \\ +\lambda_{\imath\imath}\bar{\bar{Y}}_0(\imath) + \gamma^{-2}\tilde{\bar{B}}_{1_i}(\imath)\tilde{\bar{B}}_{1_j}^T(\imath) \end{pmatrix} & (*)^T & (*)^T \\ \tilde{\bar{C}}_{1_i}(\imath)Y_0^T(\imath) + \tilde{\bar{D}}_{12}(\imath)\mathcal{C}_{0_j}(\imath) & -I & (*)^T \\ \mathcal{J}_0^T(\imath) & 0 & -\mathcal{Y}_0(\imath) \end{pmatrix} \quad (11.36)$$

$$\Psi_{22_{ij}}(\imath) = \begin{pmatrix} \begin{pmatrix} A_i^T(\imath)X_0^T(\imath) + X_0(\imath)A_i(\imath) \\ +\mathcal{B}_{0_i}(\imath)C_{2_j}(\imath) + C_{2_i}^T(\imath)\mathcal{B}_{0_j}^T(\imath) \\ +\tilde{\bar{C}}_{1_i}^T(\imath)\tilde{\bar{C}}_{1_j}(\imath) + \sum_{k=1}^s \lambda_{\imath k}\bar{\bar{X}}_0(k) \end{pmatrix} & (*)^T \\ \tilde{\bar{B}}_{1_i}^T(\imath)X_0^T(\imath) + \tilde{\bar{D}}_{21_i}^T(\imath)\mathcal{B}_{0_j}^T(\imath) & -\gamma^2 I \end{pmatrix} \quad (11.37)$$

with $\bar{\bar{X}}_0(k) = \frac{X_0(k)+X_0^T(k)}{2}$, $\bar{\bar{Y}}_0(\imath) = \frac{Y_0(\imath)+Y_0^T(\imath)}{2}$,

$$\mathcal{J}_0(\imath) = \left[\sqrt{\lambda_{1\imath}}\bar{\bar{Y}}_0(\imath) \ \cdots \ \sqrt{\lambda_{(\imath-1)\imath}}\bar{\bar{Y}}_0(\imath) \ \sqrt{\lambda_{(\imath+1)\imath}}\bar{\bar{Y}}_0(\imath) \ \cdots \ \sqrt{\lambda_{s\imath}}\bar{\bar{Y}}_0(\imath)\right]$$

$$\mathcal{Y}_0(\imath) = diag\left\{\bar{\bar{Y}}_0(1), \cdots, \bar{\bar{Y}}_0(\imath-1), \bar{\bar{Y}}_0(\imath+1), \cdots, \bar{\bar{Y}}_0(s)\right\}$$

$$\tilde{\bar{B}}_{1_i}(\imath) = [\delta(\imath)I \ I \ 0 \ B_{1_i}(\imath) \ 0]$$

$$\tilde{\bar{C}}_{1_i}(\imath) = \left[\tfrac{\gamma\bar{\rho}(\imath)}{\delta(\imath)}\bar{H}_{1_i}^T(\imath) \ \tfrac{\gamma\bar{\rho}(\imath)}{\delta(\imath)}\bar{H}_{5_i}^T(\imath) \ \sqrt{2}\bar{\aleph}(\imath)\bar{\rho}(\imath)\bar{H}_{4_i}^T(\imath) \ \sqrt{2}\bar{\aleph}(\imath)C_{1_i}^T(\imath)\right]^T$$

$$\tilde{\bar{D}}_{12}(\imath) = \left[0 \ 0 \ 0 \ -\sqrt{2}\bar{\aleph}(\imath)I\right]^T$$

$$\tilde{\bar{D}}_{21_i}(\imath) = [0 \ 0 \ \delta(\imath)I \ D_{21_i}(\imath) \ I]$$

$$\bar{\aleph}(\imath) = \left(1 + \bar{\rho}^2(\imath)\sum_{i=1}^r\sum_{j=1}^r\left[\|\bar{H}_{2_i}^T(\imath)\bar{H}_{2_j}(\imath)\| + \|\bar{H}_{7_i}^T(\imath)\bar{H}_{7_j}(\imath)\|\right]\right)^{\frac{1}{2}},$$

then there exists a sufficiently small $\hat{\varepsilon} > 0$ such that for $\varepsilon \in (0, \hat{\varepsilon}]$, the prescribed $\mathcal{H}_\infty$ performance $\gamma > 0$ is guaranteed. Furthermore, a suitable filter is of the form (11.28) with

$$\begin{aligned} \hat{A}_{ij}(\imath,\varepsilon) &= \left[Y_\varepsilon^{-1}(\imath) - X_\varepsilon(\imath)\right]^{-1}\mathcal{M}_{0_{ij}}(\imath,\varepsilon)Y_\varepsilon^{-1}(\imath) \\ \hat{B}_i(\imath) &= \left[Y_0^{-1}(\imath) - X_0(\imath)\right]^{-1}\mathcal{B}_{0_i}(\imath) \\ \hat{C}_i(\imath) &= \mathcal{C}_{0_i}(\imath)Y_0^{-1}(\imath) \end{aligned} \quad (11.38)$$

where

$$\begin{aligned} \mathcal{M}_{0_{ij}}(\imath,\varepsilon) &= -A_i^T(\imath) - X_\varepsilon(\imath)A_i(\imath)Y_\varepsilon(\imath) - \left[Y_\varepsilon^{-1}(\imath) - X_\varepsilon(\imath)\right]\hat{B}_i(\imath)C_{2_j}(\imath)Y_\varepsilon(\imath) \\ &- \textstyle\sum_{k=1}^s \lambda_{\imath k}Y_\varepsilon^{-1}(k)Y_\varepsilon(\imath) - \tilde{\bar{C}}_{1_i}^T(\imath)\left[\tilde{\bar{C}}_{1_j}(\imath)Y_\varepsilon(\imath) + \tilde{\bar{D}}_{12}(\imath)\hat{C}_j(\imath)Y_\varepsilon(\imath)\right] \\ &-\gamma^{-2}\left\{X_\varepsilon(\imath)\tilde{\bar{B}}_{1_i}(\imath) + \left[Y_\varepsilon^{-1}(\imath) - X_\varepsilon(\imath)\right]\hat{B}_i(\imath)\tilde{\bar{D}}_{21_i}(\imath)\right\}\tilde{\bar{B}}_{1_j}^T(\imath), \end{aligned} \quad (11.39)$$

$$X_\varepsilon(\imath) = \left\{X_0(\imath) + \varepsilon\tilde{X}(\imath)\right\} E_\varepsilon \tag{11.40}$$

and

$$Y_\varepsilon^{-1}(\imath) = \left\{Y_0^{-1}(\imath) + \varepsilon N_\varepsilon(\imath)\right\} E_\varepsilon \tag{11.41}$$

with $\tilde{X}(\imath) = D\Big(X_0^T(\imath) - X_0(\imath)\Big)$ *and* $N_\varepsilon(\imath) = D\Big((Y_0^{-1}(\imath))^T - Y_0^{-1}(\imath)\Big)$.

Proof: It can be shown by employing the same technique used in the proof for Theorem 18. ■

11.2 Example

Consider the tunnel diode circuit shown in Figure 4.4 where the tunnel diode is characterized by

$$i_D(t) = 0.01v_D(t) + \alpha v_D^3(t).$$

where α is the characteristic parameter. Assuming that the inductance, L, is the parasitic parameter and letting $x_1(t) = v_C(t)$ and $x_2(t) = i_L(t)$ as the state variables, the circuit is governed by the following state equations:

$$\begin{aligned} C\dot{x}_1(t) &= -0.01x_1(t) - \alpha x_1^3(t) + x_2(t) \\ L\dot{x}_2(t) &= -x_1(t) - Rx_2(t) + 0.1w_2(t) \\ y(t) &= Jx(t) + 0.1w_1(t) \\ z(t) &= \begin{bmatrix} x_1(t) \\ x_2(t) \end{bmatrix} \end{aligned} \tag{11.42}$$

where $w(t)$ is the disturbance noise input, $y(t)$ is the measurement output, $z(t)$ is the state to be estimated and J is the sensor matrix. Note that the variables $x_1(t)$ and $x_2(t)$ are the deviation variables (variables deviate from the desired trajectories). The parameters in the circuit are given as follows: $C = 100\ mF$, $R = 10\ \Omega$ and $L = \varepsilon\ H$. Suppose that this system is aggregated into 3 modes as shown in Table 11.1 with the nominal transition probability

Table 11.1. System Terminology.

Mode $\imath$	$\alpha(\imath) \pm \Delta\alpha(\imath)$
1	0.04 ±10%
2	0.05 ±10%
3	0.06 ±10%

matrix that relates the three operation modes

$$P_{\imath k} = \begin{bmatrix} 0.67 & 0.17 & 0.16 \\ 0.30 & 0.47 & 0.23 \\ 0.26 & 0.10 & 0.64 \end{bmatrix}.$$

With these parameters, (11.42) can be rewritten as

$$\begin{aligned} \dot{x}_1(t) &= -0.1x_1(t) - \left(\tfrac{[\alpha(\imath)+\Delta\alpha(\imath)]}{C} x_1^2(t)\right) \cdot x_1(t) + 10x_2(t) \\ \varepsilon\dot{x}_2(t) &= -x_1(t) - 10x_2(t) + 0.1w_2(t) \\ y(t) &= Jx(t) + 0.1w_1(t) \\ z(t) &= \begin{bmatrix} x_1(t) \\ x_2(t) \end{bmatrix}. \end{aligned} \tag{11.43}$$

For the sake of simplicity, we will use as few rules as possible. Assuming that $|x_1(t)| \le 3$, the nonlinear network system (11.43) can be approximated by the following TS fuzzy model:

Plant Rule 1: IF $x_1(t)$ is $M_1(x_1(t))$ THEN

$$\begin{aligned} E_\varepsilon \dot{x}(t) &= [A_1(\imath) + \Delta A_1(\imath)]x(t) + B_1(\imath)w(t), \quad x(0) = 0, \\ z(t) &= C_1(\imath)x(t), \\ y(t) &= C_2(\imath)x(t) + D_{21}(\imath)w(t). \end{aligned}$$

Plant Rule 2: IF $x_1(t)$ is $M_2(x_1(t))$ THEN

$$\begin{aligned} E_\varepsilon \dot{x}(t) &= [A_2(\imath) + \Delta A_2(\imath)]x(t) + B_1(\imath)w(t), \quad x(0) = 0, \\ z(t) &= C_1(\imath)x(t), \\ y(t) &= C_2(\imath)x(t) + D_{21}(\imath)w(t) \end{aligned}$$

where

$$A_1(1) = \begin{bmatrix} -0.1 & 10 \\ -1 & -10 \end{bmatrix}, \quad A_2(1) = \begin{bmatrix} -3.7 & 10 \\ -1 & -10 \end{bmatrix},$$

$$A_1(2) = \begin{bmatrix} -0.1 & 10 \\ -1 & -10 \end{bmatrix}, \quad A_2(2) = \begin{bmatrix} -4.6 & 10 \\ -1 & -10 \end{bmatrix},$$

$$A_1(3) = \begin{bmatrix} -0.1 & 10 \\ -1 & -10 \end{bmatrix}, \quad A_2(3) = \begin{bmatrix} -5.5 & 10 \\ -1 & -10 \end{bmatrix},$$

$$B_1(1) = B_1(2) = B_1(3) = \begin{bmatrix} 0 & 0 \\ 0 & 0.1 \end{bmatrix}, \quad C_1(1) = C_1(2) = C_1(3) = \begin{bmatrix} 1 & 0 \\ 0 & 1 \end{bmatrix},$$

$$C_2(1) = C_2(2) = C_2(3) = J, \quad D_{21}(1) = D_{21}(2) = D_{21}(3) = \begin{bmatrix} 0.1 & 0 \end{bmatrix},$$

$$\Delta A_1(\imath) = F(x(t), \imath, t)H_{1_1}(\imath), \quad A_2(\imath) = F(x(t), \imath, t)H_{1_2}(\imath) \text{ and } E_\varepsilon = \begin{bmatrix} 1 & 0 \\ 0 & \varepsilon \end{bmatrix}.$$

Now, by assuming that $\|F(x(t), \imath, t)\| \le \rho(\imath) = 1$, we have

$$H_{1_1}(1) = \begin{bmatrix} 0 & 0 \\ 0 & 0 \end{bmatrix}, \; H_{1_2}(1) = \begin{bmatrix} -0.36 & 0 \\ 0 & 0 \end{bmatrix},$$

$$H_{1_1}(2) = \begin{bmatrix} 0 & 0 \\ 0 & 0 \end{bmatrix}, \; H_{1_2}(2) = \begin{bmatrix} -0.45 & 0 \\ 0 & 0 \end{bmatrix},$$

$$H_{1_1}(3) = \begin{bmatrix} 0 & 0 \\ 0 & 0 \end{bmatrix} \text{ and } H_{1_2}(3) = \begin{bmatrix} -0.54 & 0 \\ 0 & 0 \end{bmatrix}.$$

Note that the plot of the membership function Rules 1 and 2 is the same as in Figure 4.5. By employing the results given in Lemma 11.1 and the Matlab LMI solver [138], it is easy to realize that $\varepsilon < 0.006$ for the fuzzy filter design in Case I and $\varepsilon < 0.008$ for the fuzzy filter design in Case II, the LMIs become ill-conditioned and the Matlab LMI solver yields the error message, "Rank Deficient".

Case I-$\nu(t)$ is available for feedback

In this case, $x_1(t) = \nu(t)$ is assumed to be available for feedback; for instance, $J = [1 \;\; 0]$. This implies that μ_i is available for feedback. Using the LMI optimization algorithm and Theorem 19 with $\varepsilon = 0.005$, $\gamma = 0.1$ and $\delta(1) = \delta(2) = \delta(3) = 1$, we obtain

$$X_0(1) = \begin{bmatrix} 0.3035 & 0.2990 \\ 0 & 1.8550 \end{bmatrix}, \qquad Y_0(1) = \begin{bmatrix} 16.1939 & -1.1293 \\ 0 & 1.8488 \end{bmatrix},$$

$$\hat{A}_{11}(1,\varepsilon) = \begin{bmatrix} -0.3401 & -0.2938 \\ -13.7069 & -4.9054 \end{bmatrix}, \; \hat{A}_{12}(1,\varepsilon) = \begin{bmatrix} -0.3401 & -0.2938 \\ -13.7069 & -4.9054 \end{bmatrix},$$

$$\hat{A}_{21}(1,\varepsilon) = \begin{bmatrix} -0.3208 & -0.2937 \\ -13.9812 & -4.9054 \end{bmatrix}, \; \hat{A}_{22}(1,\varepsilon) = \begin{bmatrix} -0.3208 & -0.2937 \\ -13.9812 & -4.9054 \end{bmatrix},$$

$$\hat{B}_1(1) = \begin{bmatrix} 1.8678 \\ 3.4400 \end{bmatrix}, \qquad \hat{B}_2(1) = \begin{bmatrix} 1.6911 \\ 3.2380 \end{bmatrix},$$

$$\hat{C}_1(1) = \begin{bmatrix} 4.2343 & -0.5409 \end{bmatrix}, \qquad \hat{C}_2(1) = \begin{bmatrix} -2.5958 & -0.5409 \end{bmatrix},$$

$$X_0(2) = \begin{bmatrix} 0.3092 & 0.3012 \\ 0 & 1.8555 \end{bmatrix}, \qquad Y_0(2) = \begin{bmatrix} 13.8969 & -0.9390 \\ 0 & 1.8359 \end{bmatrix},$$

$$\hat{A}_{11}(2,\varepsilon) = \begin{bmatrix} -0.3663 & -0.3076 \\ -19.8376 & -4.8670 \end{bmatrix}, \; \hat{A}_{12}(2,\varepsilon) = \begin{bmatrix} -0.3663 & -0.3076 \\ -19.8376 & -4.8670 \end{bmatrix},$$

$$\hat{A}_{21}(2,\varepsilon) = \begin{bmatrix} -0.3013 & -0.3074 \\ -15.8345 & -4.8670 \end{bmatrix}, \; \hat{A}_{22}(2,\varepsilon) = \begin{bmatrix} -0.3013 & -0.3074 \\ -15.8345 & -4.8670 \end{bmatrix},$$

$$\hat{B}_1(2) = \begin{bmatrix} 1.9022 \\ 3.9659 \end{bmatrix}, \qquad \hat{B}_2(2) = \begin{bmatrix} 1.6800 \\ 3.6669 \end{bmatrix},$$

$$\hat{C}_1(2) = \begin{bmatrix} -3.6022 & -0.5447 \end{bmatrix}, \qquad \hat{C}_2(2) = \begin{bmatrix} 2.2664 & -0.5447 \end{bmatrix},$$

$$X_0(3) = \begin{bmatrix} 0.3123 & 0.3025 \\ 0 & 1.8551 \end{bmatrix}, \qquad Y_0(3) = \begin{bmatrix} 16.0148 & -1.0782 \\ 0 & 1.8597 \end{bmatrix},$$

$$\hat{A}_{11}(3,\varepsilon) = \begin{bmatrix} -0.1620 & -0.2987 \\ -11.3244 & -4.9310 \end{bmatrix}, \quad \hat{A}_{12}(3,\varepsilon) = \begin{bmatrix} -0.1620 & -0.2987 \\ -11.3244 & -4.9310 \end{bmatrix},$$

$$\hat{A}_{21}(3,\varepsilon) = \begin{bmatrix} -0.2699 & -0.2985 \\ -15.4293 & -4.9310 \end{bmatrix}, \quad \hat{A}_{22}(3,\varepsilon) = \begin{bmatrix} -0.2699 & -0.2985 \\ -15.4293 & -4.9310 \end{bmatrix},$$

$$\hat{B}_1(3) = \begin{bmatrix} 1.8058 \\ 4.7164 \end{bmatrix}, \qquad \hat{B}_2(3) = \begin{bmatrix} 1.5572 \\ 4.3053 \end{bmatrix},$$

$$\hat{C}_1(3) = \begin{bmatrix} -2.1601 & -0.5377 \end{bmatrix}, \qquad \hat{C}_2(3) = \begin{bmatrix} 2.2621 & -0.5377 \end{bmatrix}.$$

The resulting fuzzy filter is

$$\begin{aligned} E_\varepsilon \dot{\hat{x}}(t) &= \textstyle\sum_{i=1}^{2} \sum_{j=1}^{2} \mu_i \mu_j \hat{A}_{ij}(\imath,\varepsilon)\hat{x}(t) + \sum_{i=1}^{2} \mu_i \hat{B}_i(\imath) y(t) \\ \hat{z}(t) &= \textstyle\sum_{i=1}^{2} \mu_i \hat{C}_i(\imath)\hat{x}(t) \end{aligned} \tag{11.44}$$

where

$$\mu_1 = M_1(x_1(t)) \text{ and } \mu_2 = M_2(x_1(t)).$$

Case II-$\nu(t)$ is unavailable for feedback

In this case, $x_1(t) = \nu(t)$ is assumed to be unavailable for feedback; for instance, $J = [0 \;\; 1]$. This implies that μ_i is unavailable for feedback. Using the LMI optimization algorithm and Theorem 20 with $\varepsilon = 0.005$, $\gamma = 0.1$ and $\delta(1) = \delta(2) = \delta(3) = 1$, we obtain

$$X_0(1) = \begin{bmatrix} 0.1716 & 0.3248 \\ 0 & 3.2335 \end{bmatrix}, \qquad Y_0(1) = \begin{bmatrix} 51.1382 & -3.3803 \\ 0 & 5.8339 \end{bmatrix},$$

$$\hat{A}_{11}(1,\varepsilon) = \begin{bmatrix} -1.1157 & -0.5675 \\ -20.1035 & -5.7388 \end{bmatrix}, \quad \hat{A}_{12}(1,\varepsilon) = \begin{bmatrix} -1.1157 & -0.5675 \\ -20.1035 & -5.7388 \end{bmatrix},$$

$$\hat{A}_{21}(1,\varepsilon) = \begin{bmatrix} -1.7859 & -0.5704 \\ -25.0273 & -5.5705 \end{bmatrix}, \quad \hat{A}_{22}(1,\varepsilon) = \begin{bmatrix} -1.7859 & -0.5704 \\ -25.0273 & -5.5705 \end{bmatrix},$$

$$\hat{B}_1(1) = \begin{bmatrix} -1.5428 \\ 1.7610 \end{bmatrix}, \qquad \hat{B}_2(1) = \begin{bmatrix} -1.5930 \\ 4.5653 \end{bmatrix},$$

$$\hat{C}_1(1) = \begin{bmatrix} -2.6689 & -0.1714 \end{bmatrix}, \qquad \hat{C}_2(1) = \begin{bmatrix} -3.3029 & -0.1714 \end{bmatrix},$$

$$X_0(2) = \begin{bmatrix} 0.1716 & 0.3248 \\ 0 & 3.2336 \end{bmatrix}, \qquad Y_0(2) = \begin{bmatrix} 43.9190 & -2.7754 \\ 0 & 5.7957 \end{bmatrix},$$

$$\hat{A}_{11}(2,\varepsilon) = \begin{bmatrix} -3.5597 & -0.5789 \\ -21.3510 & -5.7661 \end{bmatrix}, \quad \hat{A}_{12}(2,\varepsilon) = \begin{bmatrix} -3.5597 & -0.5789 \\ -21.3510 & -5.7661 \end{bmatrix},$$

$$\hat{A}_{21}(2,\varepsilon) = \begin{bmatrix} -0.6895 & -0.5829 \\ -17.9062 & -5.5030 \end{bmatrix}, \quad \hat{A}_{22}(2,\varepsilon) = \begin{bmatrix} -0.6895 & -0.5829 \\ -17.9062 & -5.5030 \end{bmatrix},$$

$$\hat{B}_1(2) = \begin{bmatrix} -1.5778 \\ 1.1687 \end{bmatrix}, \quad \hat{B}_2(2) = \begin{bmatrix} -1.6454 \\ 5.5538 \end{bmatrix},$$

$$\hat{C}_1(2) = \begin{bmatrix} 1.1729 & -0.1725 \end{bmatrix}, \quad \hat{C}_2(2) = \begin{bmatrix} -1.1600 & -0.1725 \end{bmatrix},$$

$$X_0(3) = \begin{bmatrix} 0.1714 & 0.3249 \\ 0 & 3.2331 \end{bmatrix}, \quad Y_0(3) = \begin{bmatrix} 50.8217 & -3.2190 \\ 0 & 5.8776 \end{bmatrix},$$

$$\hat{A}_{11}(3,\varepsilon) = \begin{bmatrix} -3.4814 & -0.5690 \\ -20.9458 & -5.7828 \end{bmatrix}, \quad \hat{A}_{12}(3,\varepsilon) = \begin{bmatrix} -3.4814 & -0.5690 \\ -20.9458 & -5.7828 \end{bmatrix},$$

$$\hat{A}_{21}(3,\varepsilon) = \begin{bmatrix} -1.6192 & -0.5739 \\ -22.3779 & -5.4611 \end{bmatrix}, \quad \hat{A}_{22}(3,\varepsilon) = \begin{bmatrix} -1.6192 & -0.5739 \\ -22.3779 & -5.4611 \end{bmatrix},$$

$$\hat{B}_1(3) = \begin{bmatrix} -1.5509 \\ 1.1672 \end{bmatrix}, \quad \hat{B}_2(3) = \begin{bmatrix} -1.6346 \\ 6.5293 \end{bmatrix},$$

$$\hat{C}_1(3) = \begin{bmatrix} 1.1648 & -0.1701 \end{bmatrix}, \quad \hat{C}_2(3) = \begin{bmatrix} -2.9627 & -0.1701 \end{bmatrix}.$$

The resulting fuzzy filter is

$$\begin{aligned} E_\varepsilon \dot{\hat{x}}(t) &= \textstyle\sum_{i=1}^2 \sum_{j=1}^2 \hat{\mu}_i \hat{\mu}_j \hat{A}_{ij}(\imath,\varepsilon)\hat{x}(t) + \sum_{i=1}^2 \hat{\mu}_i \hat{B}_i(\imath) y(t) \\ \hat{z}(t) &= \textstyle\sum_{i=1}^2 \hat{\mu}_i \hat{C}_i(\imath)\hat{x}(t) \end{aligned} \tag{11.45}$$

where

$$\hat{\mu}_1 = M_1(\hat{x}_1(t)) \text{ and } \hat{\mu}_2 = M_2(\hat{x}_1(t)).$$

Remark 14. Figures 11.1(a)–11.1(b), respectively, show the responses of $x_1(t)$ and $x_2(t)$ in Cases I and II. Figure 11.2 shows the result of the changing between modes during the simulation with the initial mode 2 and $\varepsilon = 0.005$. The disturbance input signal, $w(t)$, which was used during the simulation is the rectangular signal (magnitude 0.9 and frequency 0.5 Hz). The ratio of the filter error energy to the disturbance input noise energy obtained by using the $\mathcal{H}_\infty$ fuzzy filter in Case I and Case II is depicted in Figure 11.3. The ratio of the regulated output energy to the disturbance input noise energy tends to a constant value which is about 2.4×10^{-4} for the fuzzy filter in Case I and 2.8×10^{-4} for the fuzzy filter in Case II . So $\gamma = \sqrt{2.4 \times 10^{-4}} = 0.015$ for the fuzzy filter in Case I and $\gamma = \sqrt{2.8 \times 10^{-4}} = 0.017$ for the fuzzy filter in

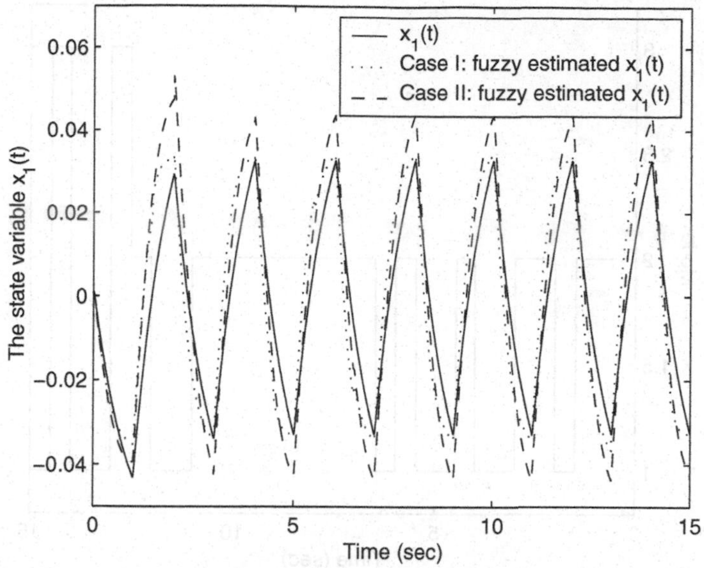

(a) The histories of $x_1(t)$

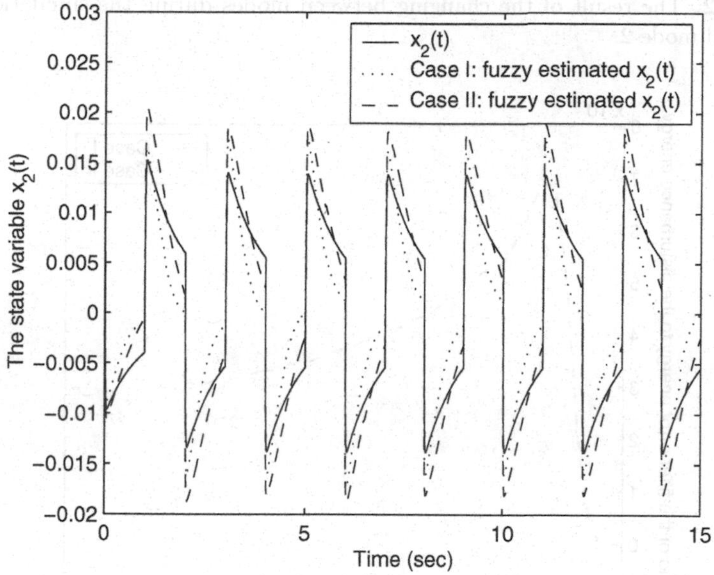

(b) The histories of $x_2(t)$

Fig. 11.1. The histories of $x_1(t)$ and $x_2(t)$ in Cases I and II.

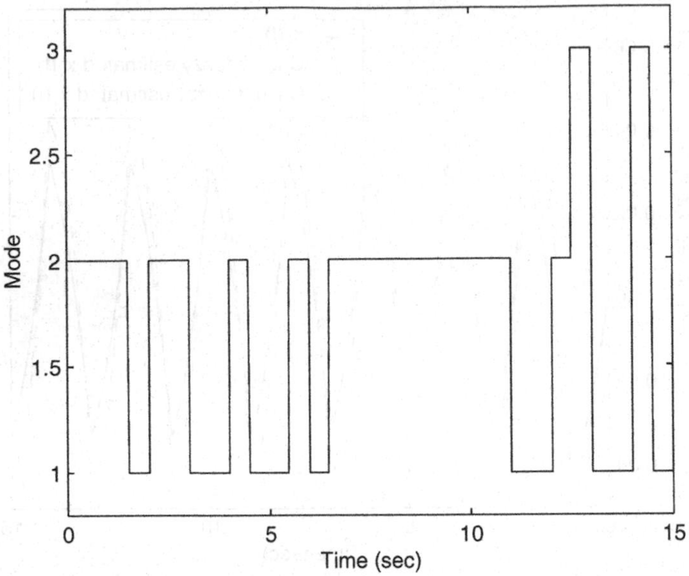

Fig. 11.2. The result of the changing between modes during the simulation with the initial mode 2.

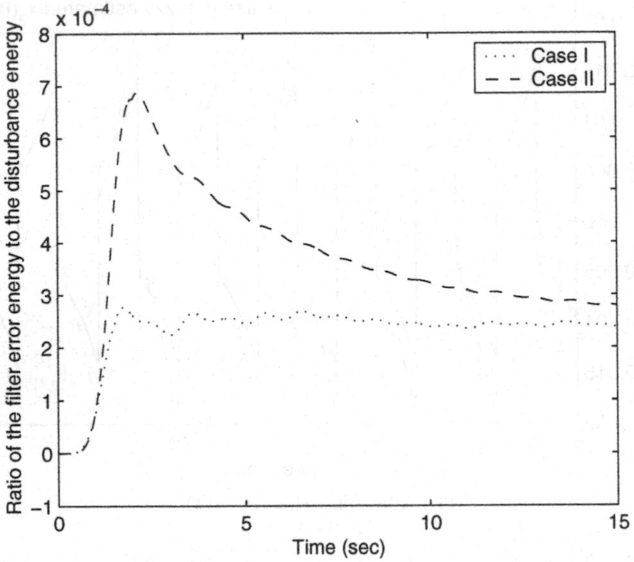

Fig. 11.3. The ratio of the filter error energy to the disturbance noise energy: $\left(\frac{\int_0^{T_f}(z(t)-\hat{z}(t))^T(z(t)-\hat{z}(t))dt}{\int_0^{T_f} w^T(t)w(t)dt}\right)$.

Table 11.2. The performance index γ of the system with different values of ε.

	The performance index γ	
ε	Output-feedback in Case I	Output-feedback in Case II
0.005	0.015	0.017
0.01	0.071	0.090
0.02	0.086	> 0.1
0.03	0.098	> 0.1
0.04	> 0.1	> 0.1

Case II which are both less than the prescribed value 0.1. *Finally, Table 11.2 shows the performance index,* γ, *for different values of* ε. *From Table 11.2, one can see that the maximum value of* ε *which guarantees the* $\mathcal{L}_2$*-gain of the mapping from the exogenous input noise to the filter error output being less than the prescribed value, is* 0.03, *i.e.,* $\varepsilon \in (0, 0.03]$ *for the fuzzy filter in Case I and* 0.01, *i.e.,* $\varepsilon \in (0, 0.01]$ *for the fuzzy filter in Case II.* □

11.3 Conclusion

This chapter has developed a new framework for designing a robust fuzzy filter for a TS singularly perturbed fuzzy system with MJs and parametric uncertainties. Sufficient conditions for the existence of the robust $\mathcal{H}_\infty$ fuzzy filter have been derived in terms of a family of ε-independent LMIs. The proposed fuzzy filter has guaranteed the $\mathcal{H}_\infty$ performance requirements. A numerical simulation example has been also presented to illustrate the theory development.

A

Proof

Proof of Theorem 3.1

Using Assumption 3.1, the closed-loop fuzzy system (3.5) can be expressed as follows:

$$\dot{x}(t) = \sum_{i=1}^{r}\sum_{j=1}^{r}\mu_i\mu_j\Big([A_i + B_{2_i}K_j]x(t) + \tilde{B}_{1_i}\tilde{w}(t)\Big) \tag{A.1}$$

where

$$\tilde{B}_{1_i} = \begin{bmatrix}\delta I & I & \delta I & B_{1_i}\end{bmatrix},$$

and the disturbance $\tilde{w}(t)$ is

$$\tilde{w}(t) = \begin{bmatrix} \frac{1}{\delta}F(x(t),t)H_{1_i}x(t) \\ F(x(t),t)H_{2_i}w(t) \\ \frac{1}{\delta}F(x(t),t)H_{3_i}K_jx(t) \\ w(t) \end{bmatrix}. \tag{A.2}$$

Let consider a Lyapunov function

$$V(x(t)) = \gamma x^T(t)Qx(t)$$

where $Q = P^{-1}$. Differentiate $V(x(t))$ along the closed-loop system (A.1) yields

$$\begin{aligned}\dot{V}(x(t)) &= \gamma\dot{x}^T(t)Qx(t) + \gamma x^T(t)Q\dot{x}(t) = \\ \sum_{i=1}^{r}\sum_{j=1}^{r}\mu_i\mu_j\Big(&\gamma x^T(t)(A_i + B_{2_i}K_j)^TQx(t) + \gamma x^T(t)Q(A_i + B_{2_i}K_j)x(t) \\ &+\gamma\tilde{w}^T(t)\tilde{B}_{1_i}^TQx(t) + \gamma x^T(t)Q\tilde{B}_{1_i}\tilde{w}(t)\Big).\end{aligned} \tag{A.3}$$

Adding and subtracting $-\tilde{z}^T(t)\tilde{z}(t) + \gamma^2\sum_{i=1}^{r}\sum_{j=1}^{r}\sum_{m=1}^{r}\sum_{n=1}^{r}\mu_i\mu_j\mu_m\mu_n \times [\tilde{w}^T(t)\tilde{w}(t)]$ to and from (A.3), we get

$$\dot{V}(x(t)) = \gamma \sum_{i=1}^{r}\sum_{j=1}^{r}\sum_{m=1}^{r}\sum_{n=1}^{r} \mu_i\mu_j\mu_m\mu_n \left[x^T(t)\ \tilde{w}^T(t) \right] \times$$

$$\left(\begin{pmatrix} (A_i + B_{2_i}K_j)^T Q + Q(A_i + B_{2_i}K_j) \\ + \frac{(\tilde{C}_{1_i}+\tilde{D}_{12_i}K_j)^T(\tilde{C}_{1_m}+\tilde{D}_{12_m}K_n)}{\gamma} \end{pmatrix} \quad (*)^T \atop \tilde{B}_{1_i}^T Q \qquad\qquad -\gamma I \right) \begin{bmatrix} x(t) \\ \tilde{w}(t) \end{bmatrix}$$

$$-\tilde{z}^T(t)\tilde{z}(t) + \gamma^2 \sum_{i=1}^{r}\sum_{j=1}^{r}\sum_{m=1}^{r}\sum_{n=1}^{r} \mu_i\mu_j\mu_m\mu_n [\tilde{w}^T(t)\tilde{w}(t)] \quad \text{(A.4)}$$

where

$$\tilde{z}(t) = \sum_{i=1}^{r}\sum_{j=1}^{r} \mu_i\mu_j [\tilde{C}_{1_i} + \tilde{D}_{12_i}K_j] x(t) \quad \text{(A.5)}$$

with $\tilde{C}_{1_i} = \left[\frac{\gamma\rho}{\delta} H_{1_i}^T\ 0\ \sqrt{2}\lambda\rho H_{4_i}^T\ \sqrt{2}\lambda C_{1_i}^T \right]^T$ and $\tilde{D}_{12_i} = \left[0\ \frac{\gamma\rho}{\delta} H_{3_i}^T\ \sqrt{2}\lambda\rho H_{6_i}^T\ \sqrt{2}\lambda D_{12_i}^T \right]^T$. Pre and post multiply (3.7)-(3.8) by $\begin{pmatrix} Q & 0 & 0 \\ 0 & I & 0 \\ 0 & 0 & I \end{pmatrix}$ yields

$$\begin{pmatrix} (A_i + B_{2_i}K_i)^T Q + Q(A_i + B_{2_i}K_i) & (*)^T & (*)^T \\ \tilde{B}_{1_i}^T Q & -\gamma I & (*)^T \\ \tilde{C}_{1_i} + \tilde{D}_{12_i}K_i & 0 & -\gamma I \end{pmatrix} < 0, \quad \text{(A.6)}$$

$i = 1, 2, \cdots, r$, and

$$\left\{ \begin{pmatrix} (A_i + B_{2_i}K_j)^T Q + Q(A_i + B_{2_i}K_j) & (*)^T & (*)^T \\ \tilde{B}_{1_i}^T Q & -\gamma I & (*)^T \\ \tilde{C}_{1_i} + \tilde{D}_{12_i}K_j & 0 & -\gamma I \end{pmatrix} \right.$$
$$\left. + \begin{pmatrix} (A_j + B_{2_j}K_i)^T Q + Q(A_j + B_{2_j}K_i) & (*)^T & (*)^T \\ \tilde{B}_{1_j}^T Q & -\gamma I & (*)^T \\ \tilde{C}_{1_j} + \tilde{D}_{12_j}K_i & 0 & -\gamma I \end{pmatrix} \right\} < 0, \quad \text{(A.7)}$$

$i < j \leq r$, respectively. Applying the Schur complement on (A.6)-(A.7) and rearranging them, then we have

$$\begin{pmatrix} \begin{pmatrix} (A_i + B_{2_i}K_i)^T Q + Q(A_i + B_{2_i}K_i) \\ + \frac{(\tilde{C}_{1_i}+\tilde{D}_{12_i}K_i)^T(\tilde{C}_{1_i}+\tilde{D}_{12_i}K_i)}{\gamma} \end{pmatrix} & (*)^T \\ \tilde{B}_{1_i}^T Q & -\gamma I \end{pmatrix} < 0, \quad \text{(A.8)}$$

$i = 1, 2, \cdots, r$, and

$$\left\{\left(\begin{array}{cc}\left(\begin{array}{c}(A_i + B_{2_i}K_j)^T Q + Q(A_i + B_{2_i}K_j) \\ + \frac{(\tilde{C}_{1_i} + \tilde{D}_{12_i}K_j)^T(\tilde{C}_{1_i} + \tilde{D}_{12_i}K_j)}{\gamma}\end{array}\right) & (*)^T \\ \tilde{B}_{1_i}^T Q & -\gamma I\end{array}\right) + \left(\begin{array}{cc}\left(\begin{array}{c}(A_j + B_{2_j}K_i)^T Q + Q(A_j + B_{2_j}K_i) \\ + \frac{(\tilde{C}_{1_j} + \tilde{D}_{12_j}K_i)^T(\tilde{C}_{1_j} + \tilde{D}_{12_j}K_i)}{\gamma}\end{array}\right) & (*)^T \\ \tilde{B}_{1_j}^T Q & -\gamma I\end{array}\right)\right\} < 0, \quad \text{(A.9)}$$

$i < j \leq r$, respectively. Using (A.8)-(A.9) and the fact that

$$\sum_{i=1}^{r}\sum_{j=1}^{r}\sum_{m=1}^{r}\sum_{n=1}^{r}\mu_i\mu_j\mu_m\mu_n M_{ij}^T N_{mn} \leq \frac{1}{2}\sum_{i=1}^{r}\sum_{j=1}^{r}\mu_i\mu_j[M_{ij}^T M_{ij} + N_{ij}N_{ij}^T], \quad \text{(A.10)}$$

it is obvious that we have

$$\left(\begin{array}{cc}\left(\begin{array}{c}(A_i + B_{2_i}K_j)^T Q + Q(A_i + B_{2_i}K_j) \\ + \frac{(\tilde{C}_{1_i} + \tilde{D}_{12_i}K_j)^T(\tilde{C}_{1_i} + \tilde{D}_{12_i}K_j)}{\gamma}\end{array}\right) & (*)^T \\ \tilde{B}_{1_i}^T Q & -\gamma I\end{array}\right) < 0 \quad \text{(A.11)}$$

where $i, j = 1, 2, \cdots, r$. Since (A.11) is less than zero and the fact that $\mu_i \geq 0$ and $\sum_{i=1}^{r}\mu_i = 1$, then (A.4) becomes

$$\dot{V}(x(t)) \leq -\tilde{z}^T(t)\tilde{z}(t) + \gamma^2\sum_{i=1}^{r}\sum_{j=1}^{r}\sum_{m=1}^{r}\sum_{n=1}^{r}\mu_i\mu_j\mu_m\mu_n[\tilde{w}^T(t)\tilde{w}(t)]. \quad \text{(A.12)}$$

Integrate both sides of (A.12) yields

$$\int_0^{T_f}\dot{V}(x(t))dt \leq \int_0^{T_f}\Big[-\tilde{z}^T(t)\tilde{z}(t) + \gamma^2\sum_{i=1}^{r}\sum_{j=1}^{r}\sum_{m=1}^{r}\sum_{n=1}^{r}\mu_i\mu_j\mu_m\mu_n \times \tilde{w}^T(t)\tilde{w}(t)\Big]dt$$

$$V(x(T_f)) - V(x(0)) \leq \int_0^{T_f}\Big[-\tilde{z}^T(t)\tilde{z}(t) + \gamma^2\sum_{i=1}^{r}\sum_{j=1}^{r}\sum_{m=1}^{r}\sum_{n=1}^{r}\mu_i\mu_j\mu_m\mu_n \times \tilde{w}^T(t)\tilde{w}(t)\Big]dt.$$

Using the fact that $x(0) = 0$ and $V(x(T_f)) \geq 0$ for all $T_f \neq 0$, we get

$$\int_0^{T_f}\tilde{z}^T(t)\tilde{z}(t)dt \leq \gamma^2\left[\int_0^{T_f}\sum_{i=1}^{r}\sum_{j=1}^{r}\sum_{m=1}^{r}\sum_{n=1}^{r}\mu_i\mu_j\mu_m\mu_n[\tilde{w}^T(t)\tilde{w}(t)]dt\right]. \quad \text{(A.13)}$$

Putting $\tilde{z}(t)$ and $\tilde{w}(t)$ respectively given in (A.5) and (A.2) into (A.13) and using the fact that $\|F(x(t),t)\| \leq \rho$, $\lambda^2 = \left(1+\rho^2 \sum_{i=1}^r \sum_{j=1}^r [\|H_{2_i}^T H_{2_j}\|]\right)$ and (A.10), we have

$$\int_0^{T_f} \sum_{i=1}^r \sum_{j=1}^r \mu_i \mu_j \Big(2\lambda^2 x^T(t)[C_{1_i} + D_{12_i} K_j]^T [C_{1_i} + D_{12_i} K_j] x(t)$$
$$+2\lambda^2 \rho^2 x^T(t)[H_{4_i} + H_{6_i} K_j]^T [H_{4_i} + H_{6_i} K_j] x(t)\Big)\, dt$$
$$\leq \gamma^2 \lambda^2 \left[\int_0^{T_f} w^T(t) w(t)\, dt\right]. \tag{A.14}$$

Adding and subtracting

$$\lambda^2 z^T(t) z(t) = \lambda^2 \sum_{i=1}^r \sum_{j=1}^r \mu_i \mu_j \Big(x^T(t)\big[C_{1_i} + F(x(t),t) H_{4_i} + D_{12_i} K_j$$
$$+F(x(t),t) H_{6_i} K_j\big]^T \big[C_{1_i} + F(x(t),t) H_{4_i} + D_{12_i} K_j$$
$$F(x(t),t) H_{6_i} K_j\big] x(t)\Big)$$

to and from (A.14), one obtains

$$\int_0^{T_f} \Big\{\lambda^2 z^T(t) z(t) + \sum_{i=1}^r \sum_{j=1}^r \mu_i \mu_j \Big(2\lambda^2 x^T(t)[C_{1_i} + D_{12_i} K_j]^T \times$$
$$[C_{1_i} + D_{12_i} K_j] x(t) + 2\lambda^2 \rho^2 x^T(t)[H_{4_i} + H_{6_i} K_j]^T [H_{4_i} + H_{6_i} K_j] x(t)$$
$$-\lambda^2 x^T(t)[C_{1_i} + F(x(t),t) H_{4_i}$$
$$+D_{12_i} K_j + F(x(t),t) H_{6_i} K_j]^T [C_{1_i} + F(x(t),t) H_{4_i} + D_{12_i} K_j$$
$$+F(x(t),t) H_{6_i} K_j] x(t)\Big)\Big\}\, dt \leq \gamma^2 \lambda^2 \left[\int_0^{T_f} w^T(t) w(t)\, dt\right]. \tag{A.15}$$

Using the triangular inequality and the fact that $\|F(x(t),t)\| \leq \rho$, we have

$$\lambda^2 \sum_{i=1}^r \sum_{j=1}^r \mu_i \mu_j \Big(x^T(t)\left[C_{1_i} + F(x(t),t) H_{4_i} + D_{12_i} K_j + F(x(t),t) H_{6_i} K_j\right]^T$$
$$\times \left[C_{1_i} + F(x(t),t) H_{4_i} + D_{12_i} K_j + F(x(t),t) H_{6_i} K_j\right] x(t)\Big)$$
$$\leq \sum_{i=1}^r \sum_{j=1}^r \mu_i \mu_j \Big(2\lambda^2 x^T(t)\left[C_{1_i} + D_{12_i} K_j\right]^T \left[C_{1_i} + D_{12_i} K_j\right] x(t)$$
$$+2\lambda^2 \rho^2 x^T(t)\left[H_{4_i} + H_{6_i} K_j\right]^T \left[H_{4_i} + H_{6_i} K_j\right] x(t)\Big). \tag{A.16}$$

Using (A.16) on (A.15), we obtain

$$\int_0^{T_f} z^T(t)z(t) \leq \gamma^2 \int_0^{T_f} w^T(t)w(t)\,dt. \tag{A.17}$$

Hence, the inequality (3.3) holds. ■

Proof of Lemma 3.1

The state space form of the fuzzy system model (3.1) with the controller (3.13) is given by

$$\begin{aligned}\dot{\breve{x}}(t) &= \textstyle\sum_{i=1}^r \sum_{j=1}^r \mu_i\mu_j \left(A_{cl}^{ij}\breve{x}(t) + B_{cl}^{ij}\tilde{w}(t)\right)\\ \breve{z}(t) &= \textstyle\sum_{i=1}^r \sum_{j=1}^r \mu_i\mu_j C_{cl}^{ij}\breve{x}(t)\end{aligned} \tag{A.18}$$

where $\breve{x}(t) = \left[x^T(t) \;\; \hat{x}^T(t)\right]^T$ and the matrix functions A_{cl}^{ij}, B_{cl}^{ij} and C_{cl}^{ij} are defined in Lemma 1 and the disturbance is

$$\tilde{w}(t) = \begin{bmatrix} \frac{1}{\delta}F(x(t),t)H_{1_i}x(t) \\ F(x(t),t)H_{2_i}w(t) \\ \frac{1}{\delta}F(x(t),t)H_{3_i}\hat{C}_j\hat{x}(t) \\ \frac{1}{\delta}F(x(t),t)H_{5_i}x(t) \\ w(t) \\ F(x(t),t)H_{7_i}w(t) \end{bmatrix}. \tag{A.19}$$

Let choose a Lyapunov function

$$V(\breve{x}(t)) = \breve{x}^T(t)Q\breve{x}(t), \tag{A.20}$$

where $Q = P^{-1}$. Differentiate $V(\breve{x}(t))$ along the closed-loop system (A.18) yields

$$\begin{aligned}\dot{V}(\breve{x}(t)) &= \dot{\breve{x}}^T(t)Q\breve{x}(t) + \breve{x}^T(t)Q\dot{\breve{x}}(t)\\ &= \sum_{i=1}^r\sum_{j=1}^r \mu_i\mu_j\Big(\breve{x}^T(t)(A_{cl}^{ij})^TQ\breve{x}(t) + \breve{x}^T(t)QA_{cl}^{ij}\breve{x}(t)\\ &\quad +\tilde{w}^T(t)(B_{cl}^{ij})^TQ\breve{x}(t) + \breve{x}^T(t)QB_{cl}^{ij}\tilde{w}(t)\Big).\end{aligned} \tag{A.21}$$

Add and subtract

$$-\breve{z}^T(t)\breve{z}(t) + \gamma^2\sum_{i=1}^r\sum_{j=1}^r\sum_{m=1}^r\sum_{n=1}^r \mu_i\mu_j\mu_m\mu_n[\tilde{w}(t)^T\tilde{w}(t)]$$

to and from (A.21) yields

$$\dot{V}(\breve{x}(t)) = \sum_{i=1}^{r}\sum_{j=1}^{r}\sum_{m=1}^{r}\sum_{n=1}^{r} \mu_i\mu_j\mu_m\mu_n \left[\breve{x}^T(t)\ \tilde{w}^T(t)\right]$$
$$\begin{pmatrix} \begin{pmatrix} (A_{cl}^{ij})^T Q + Q A_{cl}^{ij} \\ +(C_{cl}^{ij})^T C_{cl}^{mn} \end{pmatrix} & (*)^T \\ Q B_{cl}^{ij} & -\gamma^2 I \end{pmatrix} \begin{bmatrix} \breve{x}(t) \\ \tilde{w}(t) \end{bmatrix}$$
$$-\breve{z}^T(t)\breve{z}(t) + \gamma^2 \sum_{i=1}^{r}\sum_{j=1}^{r}\sum_{m=1}^{r}\sum_{n=1}^{r} \mu_i\mu_j\mu_m\mu_n[\tilde{w}^T(t)\tilde{w}(t)]. \quad \text{(A.22)}$$

Now suppose there exits a matrix $P > 0$ such that (3.15) holds, i.e.,

$$\begin{pmatrix} A_{cl}^{ij} P + P(A_{cl}^{ij})^T & (*)^T & (*)^T \\ (B_{cl}^{ij})^T & -\gamma^2 I & (*)^T \\ C_{cl}^{ij} P & 0 & -I \end{pmatrix} < 0. \quad \text{(A.23)}$$

Pre and post multiply (A.23) by $\begin{pmatrix} Q & 0 & 0 \\ 0 & I & 0 \\ 0 & 0 & I \end{pmatrix}$ yields

$$\begin{pmatrix} (A_{cl}^{ij})^T Q + Q A_{cl}^{ij} & (*)^T & (*)^T \\ (B_{cl}^{ij})^T Q & -\gamma^2 I & (*)^T \\ C_{cl}^{ij} & 0 & -I \end{pmatrix} < 0. \quad \text{(A.24)}$$

The Schur complement of (A.24) is

$$\begin{pmatrix} (A_{cl}^{ij})^T Q + Q A_{cl}^{ij} + (C_{cl}^{ij})^T C_{cl}^{ij} & (*)^T \\ (B_{cl}^{ij})^T & -\gamma^2 I \end{pmatrix} < 0. \quad \text{(A.25)}$$

Using (A.25) and the fact in (A.10) together with the fact that $\mu_i \geq 0$ and $\sum_{i=1}^{r} \mu_i = 1$, then (A.22) becomes

$$\dot{V}(\breve{x}(t)) \leq -\breve{z}^T(t)\breve{z}(t) + \gamma^2 \sum_{i=1}^{r}\sum_{j=1}^{r}\sum_{m=1}^{r}\sum_{n=1}^{r} \mu_i\mu_j\mu_m\mu_n[\tilde{w}^T(t)\tilde{w}(t)]. \quad \text{(A.26)}$$

Integrate both sides of (A.26) yields

$$\int_0^{T_f} \dot{V}(\breve{x}(t))dt \leq \int_0^{T_f} \Big(-\breve{z}^T(t)\breve{z}(t) + \gamma^2 \sum_{i=1}^{r}\sum_{j=1}^{r}\sum_{m=1}^{r}\sum_{n=1}^{r} \mu_i\mu_j\mu_m\mu_n \times$$
$$[\tilde{w}^T(t)\tilde{w}(t)]\Big)dt$$
$$V(\breve{x}(T_f)) - V(\breve{x}(0)) \leq \int_0^{T_f} \Big(-\breve{z}^T(t)\breve{z}(t) + \gamma^2 \sum_{i=1}^{r}\sum_{j=1}^{r}\sum_{m=1}^{r}\sum_{n=1}^{r} \mu_i\mu_j\mu_m\mu_n \times$$
$$[\tilde{w}^T(t)\tilde{w}(t)]\Big)dt.$$

Using the fact that $\breve{x}(0) = 0$ and $V(\breve{x}(T_f)) > 0$ for all $T_f \neq 0$, we have

$$\int_0^{T_f} \breve{z}^T(t)\breve{z}(t)dt \leq \gamma^2 \left[\int_0^{T_f} \sum_{i=1}^{r}\sum_{j=1}^{r}\sum_{m=1}^{r}\sum_{n=1}^{r} \mu_i\mu_j\mu_m\mu_n [\tilde{w}^T(t)\tilde{w}(t)] \right] dt. \quad (A.27)$$

Putting $\breve{z}(t)$ and $\tilde{w}(t)$ respectively given in (A.18) and (A.19) into (A.27) and using the fact that $\|F(x(t),t)\| \leq \rho$, $\lambda^2 = \left(1 + \rho^2 \sum_{i=1}^{r}\sum_{j=1}^{r}\left[\|H_{2_i}^T H_{2_j}\| + \|H_{7_i}^T H_{7_j}\|\right]\right)$ and (A.10), we have

$$\int_0^{T_f} \sum_{i=1}^{r}\sum_{j=1}^{r} \mu_i\mu_j \Big(2\lambda^2 \breve{x}^T(t)[C_{1_i} \;\; D_{12_i}\hat{C}_j]^T [C_{1_i} \;\; D_{12_i}\hat{C}_j]\breve{x}(t)$$

$$+2\lambda^2\rho^2\breve{x}^T(t)[H_{4_i} \;\; H_{6_i}\hat{C}_j]^T[H_{4_i} \;\; H_{6_i}\hat{C}_j]\breve{x}(t)\Big)\, dt \leq \gamma^2\lambda^2 \left[\int_0^{T_f} w^T(t)w(t)\, dt\right]. \quad (A.28)$$

Adding and subtracting

$$\lambda^2 z^T(t)z(t) = \lambda^2 \sum_{i=1}^{r}\sum_{j=1}^{r} \mu_i\mu_j \Big(\breve{x}^T(t) \Big[C_{1_i} + F(x(t),t)H_{4_i} \;\; D_{12_i}\hat{C}_j$$
$$+F(x(t),t)H_{6_i}\hat{C}_j\Big]^T \Big[C_{1_i} + F(x(t),t)H_{4_i} \;\; D_{12_i}\hat{C}_j$$
$$+F(x(t),t)H_{6_i}\hat{C}_j\Big]\breve{x}(t)\Big)$$

to and from (A.28), one obtains

$$\int_0^{T_f} \Big\{ \lambda^2 z^T(t)z(t) + \sum_{i=1}^{r}\sum_{j=1}^{r} \mu_i\mu_j \Big(2\lambda^2\breve{x}^T(t)[C_{1_i} \;\; D_{12_i}\hat{C}_j]^T[C_{1_i} \;\; D_{12_i}\hat{C}_j]\breve{x}(t)$$
$$+2\lambda^2\rho^2\breve{x}^T(t)[H_{4_i} \;\; H_{6_i}\hat{C}_j]^T[H_{4_i} \;\; H_{6_i}\hat{C}_j]\breve{x}(t)$$
$$-\lambda^2\breve{x}^T(t)[C_{1_i} + F(x(t),t)H_{4_i} \;\; D_{12_i}\hat{C}_j + F(x(t),t)H_{6_i}\hat{C}_j]^T \times$$
$$[C_{1_i} + F(x(t),t)H_{4_i} \;\; D_{12_i}\hat{C}_j + F(x(t),t)H_{6_i}\hat{C}_j]\breve{x}(t)\Big)\Big\}\, dt$$
$$\leq \gamma^2\lambda^2\left[\int_0^{T_f} w^T(t)w(t)\, dt\right]. \quad (A.29)$$

Using the triangular inequality and the fact that $\|F(x(t),t)\| \leq \rho$, we have

$$\lambda^2 \sum_{i=1}^{r}\sum_{j=1}^{r} \mu_i\mu_j\Big(\breve{x}^T(t)\Big[C_{1_i} + F(x(t),t)H_{4_i} \quad D_{12_i}\hat{C}_j + F(x(t),t)H_{6_i}\hat{C}_j\Big]^T \times$$

$$\Big[C_{1_i} + F(x(t),t)H_{4_i} \quad D_{12_i}\hat{C}_j + F(x(t),t)H_{6_i}\hat{C}_j\Big]\breve{x}(t)\Big)$$

$$\leq \sum_{i=1}^{r}\sum_{j=1}^{r} \mu_i\mu_j\Big(2\lambda^2\breve{x}^T(t)\Big[C_{1_i} \quad D_{12_i}\hat{C}_j\Big]^T\Big[C_{1_i} \quad D_{12_i}\hat{C}_j\Big]\breve{x}(t)$$

$$+2\lambda^2\rho^2\breve{x}^T(t)\Big[H_{4_i} \quad H_{6_i}\hat{C}_j\Big]^T\Big[H_{4_i} \quad H_{6_i}\hat{C}_j\Big]\breve{x}(t)\Big). \tag{A.30}$$

Using (A.30) on (A.29), we obtain

$$\int_0^{T_f} z^T(t)z(t) \leq \gamma^2 \int_0^{T_f} w^T(t)w(t)\, dt. \tag{A.31}$$

Hence, the inequality (3.3) is guaranteed. ∎

Proof of Theorem 5.1

The closed-loop state space form of the fuzzy system model (5.1) with the controller (5.6) is given by

$$\dot{x}(t) = \sum_{i=1}^{r}\sum_{j=1}^{r} \mu_i\mu_j\Big([A_i(\imath) + B_{2_i}(\imath)K_j(\imath)]x(t) + [\Delta A_i(\imath) + \Delta B_{2_i}(\imath)K_j(\imath)]x(t)$$

$$+[B_{1_i}(\imath) + \Delta B_{1_i}(\imath)]w(t)\Big), \quad x(0) = 0, \tag{A.32}$$

or in a more compact form

$$\dot{x}(t) = \textstyle\sum_{i=1}^{r}\sum_{j=1}^{r} \mu_i\mu_j\Big([A_i(\imath) + B_{2_i}(\imath)K_j(\imath)]x(t) + \tilde{B}_{1_i}(\imath)\mathcal{R}(\imath)\tilde{w}(t)\Big) \tag{A.33}$$

where

$$\tilde{B}_{1_i}(\imath) = \big[\, I \; I \; I \; B_{1_i}(\imath)\,\big] \tag{A.34}$$

$$\tilde{w}(t) = \mathcal{R}^{-1}(\imath)\begin{bmatrix} F(x(t),\imath,t)H_{1_i}(\imath)x(t) \\ F(x(t),\imath,t)H_{2_i}(\imath)w(t) \\ F(x(t),\imath,t)H_{3_i}(\imath)K_j(\imath)x(t) \\ w(t) \end{bmatrix}. \tag{A.35}$$

Consider a Lyapunov functional candidate as follows:

$$V(x(t),\imath) = \gamma x^T(t)Q(\imath)x(t), \quad \forall\, \imath \in \mathcal{S}. \tag{A.36}$$

Note that $Q(\imath)$ is constant for each $\imath$. For this choice, we have $V(0,\imath_0) = 0$ and $V(x(t),\imath) \to \infty$ only when $\|x(t)\| \to \infty$.

Now let consider the weak infinitesimal operator $\tilde{\Delta}$ of the joint process $\{(x(t), \imath), t \geq 0\}$, which is the stochastic analog of the deterministic derivative. $\{(x(t), \imath), t \geq 0\}$ is a Markov process with infinitesimal operator given by [80],

$$
\begin{aligned}
\tilde{\Delta}V(x(t), \imath) &= \gamma \dot{x}^T(t)Q(\imath)x(t) + \gamma x^T(t)Q(\imath)\dot{x}(t) + \gamma x^T(t)\sum_{k=1}^{s} \lambda_{\imath k} Q(k)x(t) \\
&= \sum_{i=1}^{r}\sum_{j=1}^{r} \mu_i \mu_j \Big(\gamma x^T(t)Q(\imath)\left[(A_i(\imath) + B_{2_i}(\imath)K_j(\imath))\right]x(t) \\
&\quad + \gamma x^T(t)\left[A_i(\imath) + B_{2_i}(\imath)K_j(\imath)\right]^T Q(\imath)x(t) \\
&\quad + \gamma x^T(t)Q(\imath)\tilde{B}_{1_i}(\imath)\mathcal{R}(\imath)\tilde{w}(t) \\
&\quad + \gamma \tilde{w}^T(t)\mathcal{R}(\imath)\tilde{B}_{1_i}^T(\imath)Q(\imath)x(t) + \gamma x^T(t)\sum_{k=1}^{s}\lambda_{\imath k}Q(k)x(t)\Big) \qquad \text{(A.37)}
\end{aligned}
$$

Adding and subtracting

$$
-\aleph^2(\imath)z^T(t)z(t) + \gamma^2 \sum_{i=1}^{r}\sum_{j=1}^{r}\sum_{m=1}^{r}\sum_{n=1}^{r} \mu_i\mu_j\mu_m\mu_n[\tilde{w}^T(t)\mathcal{R}(\imath)\tilde{w}(t)]
$$

to and from (A.37), we get

$$
\begin{aligned}
\tilde{\Delta}V(x(t), \imath) &= -\aleph^2(\imath)z^T(t)z(t) + \gamma^2 \sum_{i=1}^{r}\sum_{j=1}^{r}\sum_{m=1}^{r}\sum_{n=1}^{r} \mu_i\mu_j\mu_m\mu_n[\tilde{w}^T(t)\mathcal{R}(\imath)\tilde{w}(t)] \\
&\quad + \aleph^2(\imath)z^T(t)z(t) + \gamma \sum_{i=1}^{r}\sum_{j=1}^{r}\sum_{m=1}^{r}\sum_{n=1}^{r} \mu_i\mu_j\mu_m\mu_n \begin{bmatrix} x(t) \\ \tilde{w}(t) \end{bmatrix}^T \times \\
&\quad \left(\begin{array}{cc} \left(\begin{array}{c} \left[A_i(\imath) + B_{2_i}(\imath)K_j(\imath)\right]^T Q(\imath) \\ + Q(\imath)\left[A_i(\imath) + B_{2_i}(\imath)K_j(\imath)\right] \\ + \sum_{k=1}^{s} \lambda_{\imath k}Q(k) \end{array} \right) & (*)^T \\ \mathcal{R}(\imath)\tilde{B}_{1_i}^T(\imath)Q(\imath) & -\gamma\mathcal{R}(\imath) \end{array} \right) \begin{bmatrix} x(t) \\ \tilde{w}(t) \end{bmatrix}. \qquad \text{(A.38)}
\end{aligned}
$$

Now let us consider the following terms:

$$
\begin{aligned}
&\gamma^2 \sum_{i=1}^{r}\sum_{j=1}^{r}\sum_{m=1}^{r}\sum_{n=1}^{r} \mu_i\mu_j\mu_m\mu_n[\tilde{w}^T(t)\mathcal{R}(\imath)\tilde{w}(t)] = \gamma^2 \sum_{i=1}^{r}\sum_{j=1}^{r}\sum_{m=1}^{r}\sum_{n=1}^{r} \mu_i\mu_j\mu_m\mu_n \\
&\times \begin{bmatrix} F(x(t), \imath, t)H_{1_i}(\imath)x(t) \\ F(x(t), \imath, t)H_{2_i}(\imath)w(t) \\ F(x(t), \imath, t)H_{3_i}(\imath)K_j(\imath)x(t) \\ w(t) \end{bmatrix}^T \mathcal{R}^{-1}(\imath) \begin{bmatrix} F(x(t), \imath, t)H_{1_m}(\imath)x(t) \\ F(x(t), \imath, t)H_{2_m}(\imath)w(t) \\ F(x(t), \imath, t)H_{3_m}(\imath)K_n(\imath)x(t) \\ w(t) \end{bmatrix} \\
&\leq \frac{\rho^2(\imath)\gamma^2}{\delta(\imath)} \sum_{i=1}^{r}\sum_{j=1}^{r}\sum_{m=1}^{r}\sum_{n=1}^{r} \mu_i\mu_j\mu_m\mu_n x^T(t)\Big\{ H_{1_i}^T(\imath)H_{1_m}(\imath) \\
&+ K_j^T(\imath)H_{3_i}^T(\imath)H_{3_m}(\imath)K_n(\imath)\Big\} x(t) + \aleph^2(\imath)\gamma^2 w^T(t)w(t) \qquad \text{(A.39)}
\end{aligned}
$$

and

$$\begin{aligned}
\aleph^2(\imath)z^T(t)z(t) &= \aleph^2(\imath)\sum_{i=1}^{r}\sum_{j=1}^{r}\sum_{m=1}^{r}\sum_{n=1}^{r}\mu_i\mu_j\mu_m\mu_n x^T(t)\Big[C_{1_i}(\imath)+ \\
&\quad F(x(t),\imath,t)H_{4_i}(\imath)+D_{12_i}(\imath)K_j(\imath)+F(x(t),\imath,t)H_{6_i}(\imath)K_j(\imath)\Big]^T \\
&\quad \times\Big[C_{1_m}(\imath)+F(x(t),\imath,t)H_{4_m}(\imath) \\
&\quad +D_{12_m}(\imath)K_n(\imath)+F(x(t),\imath,t)H_{6_m}(\imath)K_n(\imath)\Big]x(t) \\
&\leq 2\aleph^2(\imath)\sum_{i=1}^{r}\sum_{j=1}^{r}\sum_{m=1}^{r}\sum_{n=1}^{r}\mu_i\mu_j\mu_m\mu_n x^T(t)\Big\{ \\
&\quad [C_{1_i}(\imath)+D_{12_i}(\imath)K_j(\imath)]^T\,[C_{1_m}(\imath)+D_{12_m}(\imath)K_n(\imath)]+ \\
&\quad \rho^2(\imath)\,[H_{4_i}(\imath)+H_{6_i}(\imath)K_j(\imath)]^T\times \\
&\quad [H_{4_m}(\imath)+H_{6_m}(\imath)K_n(\imath)]\Big\}x(t)
\end{aligned} \tag{A.40}$$

where $\aleph(\imath)=\Big(1+\rho^2(\imath)\sum_{i=1}^{r}\sum_{j=1}^{r}[\|H_{2_i}^T(\imath)H_{2_j}(\imath)\|]\Big)^{\frac{1}{2}}$. Hence,

$$\begin{aligned}
&\gamma^2\sum_{i=1}^{r}\sum_{j=1}^{r}\sum_{m=1}^{r}\sum_{n=1}^{r}\mu_i\mu_j\mu_m\mu_n[\tilde{w}^T(t)\mathcal{R}(\imath)\tilde{w}(t)]+\aleph^2(\imath)z^T(t)z(t) \\
&\leq\sum_{i=1}^{r}\sum_{j=1}^{r}\sum_{m=1}^{r}\sum_{n=1}^{r}\mu_i\mu_j\mu_m\mu_n\Big(x^T(t)\Big[\tilde{C}_{1_i}(\imath)+\tilde{D}_{12_i}(\imath)K_j(\imath)\Big]^T\mathcal{R}^{-1}(\imath)\times \\
&\Big[\tilde{C}_{1_m}(\imath)+\tilde{D}_{12_m}(\imath)K_n(\imath)\Big]x(t)\Big)+\aleph^2(\imath)\gamma^2 w^T(t)w(t)
\end{aligned} \tag{A.41}$$

where

$$\begin{aligned}
\tilde{C}_{1_i}(\imath) &= \big[\,\gamma\rho(\imath)H_{1_i}^T(\imath)\ \sqrt{2}\aleph(\imath)\rho(\imath)H_{4_i}^T(\imath)\ 0\ \sqrt{2}\aleph(\imath)C_{1_i}^T(\imath)\,\big]^T \\
\tilde{D}_{12_i}(\imath) &= \big[\,0\ \sqrt{2}\aleph(\imath)\rho(\imath)H_{6_i}^T(\imath)\ \gamma\rho(\imath)H_{3_i}^T(\imath)\ \sqrt{2}\aleph(\imath)D_{12_i}^T(\imath)\,\big]^T.
\end{aligned}$$

Substituting (A.41) into (A.38), we have

$$\begin{aligned}
\tilde{\Delta}V(x(t),\imath) &\leq -\aleph^2(\imath)z^T(t)z(t)+\gamma^2\aleph^2(\imath)w^T(t)w(t) \\
&\quad +\gamma\sum_{i=1}^{r}\sum_{j=1}^{r}\sum_{m=1}^{r}\sum_{n=1}^{r}\mu_i\mu_j\mu_m\mu_n\begin{bmatrix}x(t)\\ \tilde{w}(t)\end{bmatrix}^T\Phi_{ijmn}(\imath)\begin{bmatrix}x(t)\\ \tilde{w}(t)\end{bmatrix}
\end{aligned} \tag{A.42}$$

where

$$\Phi_{ijmn}(\imath) = \begin{pmatrix} \begin{pmatrix} (A_i(\imath) + B_{2_i}(\imath)K_j(\imath))^T Q(\imath) \\ +Q(\imath)\Big[A_i(\imath) + B_{2_i}(\imath)K_j(\imath)\Big] \\ +\frac{1}{\gamma}\Big[\tilde{C}_{1_i}(\imath) + \tilde{D}_{12_i}(\imath)K_j(\imath)\Big]^T \times \\ \mathcal{R}^{-1}(\imath)\Big[\tilde{C}_{1_m}(\imath) + \tilde{D}_{12_m}(\imath)K_n(\imath)\Big] \\ +\sum_{k=1}^{s}\lambda_{\imath k}Q(k) \end{pmatrix} & (*)^T \\ \mathcal{R}(\imath)\tilde{B}_{1_i}^T(\imath)Q(\imath) & -\gamma\mathcal{R}(\imath) \end{pmatrix}. \tag{A.43}$$

Using the fact

$$\sum_{i=1}^{r}\sum_{j=1}^{r}\sum_{m=1}^{r}\sum_{n=1}^{r}\mu_i\mu_j\mu_m\mu_n M_{ij}^T(\imath)N_{mn}(\imath) \leq \frac{1}{2}\sum_{i=1}^{r}\sum_{j=1}^{r}\mu_i\mu_j[M_{ij}^T(\imath)M_{ij}(\imath) + N_{ij}(\imath)N_{ij}^T(\imath)],$$

we can rewrite (A.42) as follows:

$$\begin{aligned} \tilde{\Delta}V(x(t),\imath) &\leq -\aleph^2(\imath)z^T(t)z(t) + \gamma^2\aleph^2(\imath)w^T(t)w(t) \\ &\quad +\gamma\sum_{i=1}^{r}\sum_{j=1}^{r}\mu_i\mu_j\begin{bmatrix} x(t) \\ \tilde{w}(t)\end{bmatrix}^T \Phi_{ij}(\imath)\begin{bmatrix} x(t) \\ \tilde{w}(t)\end{bmatrix} \\ &= -\aleph^2(\imath)z^T(t)z(t) + \gamma^2\aleph^2(\imath)w^T(t)w(t) \\ &\quad +\gamma\sum_{i=1}^{r}\mu_i^2\begin{bmatrix} x(t) \\ \tilde{w}(t)\end{bmatrix}^T \Phi_{ii}(\imath)\begin{bmatrix} x(t) \\ \tilde{w}(t)\end{bmatrix} \\ &\quad +\gamma\sum_{i=1}^{r}\sum_{i<j}^{r}\mu_i\mu_j\begin{bmatrix} x(t) \\ \tilde{w}(t)\end{bmatrix}^T \Big(\Phi_{ij}(\imath) + \Phi_{ji}(\imath)\Big)\begin{bmatrix} x(t) \\ \tilde{w}(t)\end{bmatrix} \end{aligned} \tag{A.44}$$

where

$$\Phi_{ij}(\imath) = \begin{pmatrix} \begin{pmatrix} (A_i(\imath) + B_{2_i}(\imath)K_j(\imath))^T Q(\imath) \\ +Q(\imath)(A_i(\imath) + B_{2_i}(\imath)K_j(\imath)) \\ +\frac{1}{\gamma}\Big[\tilde{C}_{1_i}(\imath) + \tilde{D}_{12_i}(\imath)K_j(\imath)\Big]^T \times \\ \mathcal{R}^{-1}(\imath)\Big[\tilde{C}_{1_i}(\imath) + \tilde{D}_{12_i}(\imath)K_j(\imath)\Big] \\ +\sum_{k=1}^{s}\lambda_{\imath k}Q(k) \end{pmatrix} & (*)^T \\ \mathcal{R}(\imath)\tilde{B}_{1_i}^T(\imath)Q(\imath) & -\gamma\mathcal{R}(\imath) \end{pmatrix}. \tag{A.45}$$

Pre and post multiplying (A.45) by

$$\Xi(\imath) = \begin{pmatrix} P(\imath) & 0 \\ 0 & I \end{pmatrix},$$

with $P(\imath) = Q^{-1}(\imath)$, we obtain

$$\Xi(\imath)\Phi_{ij}(\imath)\Xi(\imath) = \left(\begin{pmatrix} P(\imath)A_i^T(\imath) + Y_j^T(\imath)B_{2_i}^T(\imath) \\ +A_i(\imath)P(\imath) + B_{2_i}(\imath)Y_j(\imath) \\ +\frac{1}{\gamma}\left[\tilde{C}_{1_i}(\imath)P(\imath) + \tilde{D}_{12_i}(\imath)Y_j(\imath)\right]^T \mathcal{R}^{-1}(\imath)\times \\ \left[\tilde{C}_{1_i}(\imath)P(\imath) + \tilde{D}_{12_i}(\imath)Y_j(\imath)\right] \\ +\sum_{k=1}^{s} \lambda_{\imath k}P(\imath)P^{-1}(k)P(\imath) \end{pmatrix} \quad (*)^T \\ \mathcal{R}(\imath)\tilde{B}_{1_i}^T(\imath) \quad -\gamma\mathcal{R}(\imath)\right). \tag{A.46}$$

Note that (A.46) is the Schur complement of $\Psi_{ij}(\imath)$ defined in (5.11). Using (5.9)-(5.10), we learn that

$$\Phi_{ii}(\imath) < 0 \tag{A.47}$$

$$\Phi_{ij}(\imath) + \Phi_{ji}(\imath) < 0. \tag{A.48}$$

Following from (A.44), (A.47) and (A.48), we know that

$$\tilde{\Delta}V(x(t),\imath) < -\aleph^2(\imath)z^T(t)z(t) + \gamma^2\aleph^2(\imath)w^T(t)w(t). \tag{A.49}$$

Applying the operator $\mathbf{E}[\int_0^{T_f}(\cdot)dt]$ on both sides of (A.49), we obtain

$$\mathbf{E}\left[\int_0^{T_f} \tilde{\Delta}V(x(t),\imath)dt\right] < \mathbf{E}\left[\int_0^{T_f} (-\aleph^2(\imath)z^T(t)z(t) + \gamma^2\aleph^2(\imath)w^T(t)w(t))dt\right]. \tag{A.50}$$

From the Dynkin's formula [75], it follows that

$$\mathbf{E}\left[\int_0^{T_f} \tilde{\Delta}V(x(t),\imath)dt\right] = \mathbf{E}[V(x(T_f),\imath(T_f))] - \mathbf{E}[V(x(0),\imath(0))]. \tag{A.51}$$

Substitute (A.51) into (A.50) yields

$$0 < \mathbf{E}\left[\int_0^{T_f} (-\aleph^2(\imath)z^T(t)z(t) + \gamma^2\aleph^2(\imath)w^T(t)w(t))dt\right] \\ -\mathbf{E}[V(x(T_f),\imath(T_f))] + \mathbf{E}[V(x(0),\imath(0))].$$

Using (A.49) and the fact that $V(x(0)=0,\imath(0)) = 0$ and $V(x(T_f),\imath(T_f)) > 0$, we have

$$\mathbf{E}\left[\int_0^{T_f} \left\{z^T(t)z(t) - \gamma^2 w^T(t)w(t)\right\}dt\right] < 0. \tag{A.52}$$

Hence, the inequality (5.5) holds. This completes the proof of Theorem 6. ∎

Proof of Lemma 5.1

The closed-loop state space form of the fuzzy system model (5.1) with the controller (5.24) is given by

$$\begin{aligned} \dot{\breve{x}}(t) &= \sum_{i=1}^{r}\sum_{j=1}^{r}\mu_i\mu_j\Big(A_{cl}^{ij}(\imath)\breve{x}(t) + B_{cl}^{ij}(\imath)\tilde{w}(t)\Big) \\ \breve{z}(t) &= \sum_{i=1}^{r}\sum_{j=1}^{r}\mu_i\mu_j C_{cl}^{ij}(\imath)\breve{x}(t) \end{aligned} \tag{A.53}$$

where $\breve{x}(t) = \left[x^T(t)\ \ \hat{x}^T(t)\right]^T$ and the matrix functions $A_{cl}^{ij}(\imath)$, $B_{cl}^{ij}(\imath)$ and $C_{cl}^{ij}(\imath)$ are defined in Lemma 3 and the disturbance is

$$\tilde{w}(t) = \begin{bmatrix} \frac{1}{\delta(\imath)}F(x(t),\imath,t)H_{1_i}(\imath)x(t) \\ F(x(t),\imath,t)H_{2_i}(\imath)w(t) \\ \frac{1}{\delta(\imath)}F(x(t),\imath,t)H_{3_i}(\imath)\hat{C}_j(\imath)\hat{x}(t) \\ \frac{1}{\delta(\imath)}F(x(t),\imath,t)H_{5_i}(\imath)x(t) \\ w(t) \\ F(x(t),\imath,t)H_{7_i}(\imath)w(t) \end{bmatrix}.$$

Let choose a stochastic Lyapunov function

$$V(\breve{x}(t),\imath) = \breve{x}^T(t)P(\imath)\breve{x}(t) \ \ \forall\, \imath \in \mathcal{S} \tag{A.54}$$

where $P(\imath)$ is a constant positive definite matrix for each $\imath$. For this choice, we have $V(0,\imath_0) = 0$ and $V(\breve{x}(t),\imath) \to \infty$ only when $\|\breve{x}(t)\| \to \infty$.

Consider the weak infinitesimal operator $\tilde{\Delta}$ of the joint process $\{(\breve{x}(t),\imath), t \geq 0\}$, which is the stochastic analog of the deterministic derivative. $\{(\breve{x}(t),\imath), t \geq 0\}$ is a Markov process with infinitesimal operator given by [80],

$$\begin{aligned} \tilde{\Delta}V(\breve{x}(t),\imath) &= \dot{\breve{x}}^T(t)P(\imath)\breve{x}(t) + \breve{x}^T(t)P(\imath)\dot{\breve{x}}(t) + \breve{x}^T(t)\sum_{k=1}^{s}\lambda_{\imath k}P(k)\breve{x}^T(t) \\ &= \sum_{i=1}^{r}\sum_{j=1}^{r}\mu_i\mu_j\Big(\breve{x}^T(t)(A_{cl}^{ij}(\imath))^T P(\imath)\breve{x}(t) + \breve{x}^T(t)P(\imath)A_{cl}^{ij}(\imath)\breve{x}(t) \\ &\quad + \tilde{w}^T(t)(B_{cl}^{ij}(\imath))^T P(\imath)\breve{x}(t) + \breve{x}^T(t)P(\imath)B_{cl}^{ij}(\imath)\tilde{w}(t) \\ &\quad + \breve{x}^T(t)\sum_{k=1}^{s}\lambda_{\imath k}P(k)\breve{x}^T(t)\Big). \end{aligned} \tag{A.55}$$

Adding and subtracting

$$-\aleph^2(\imath)z^T(t)z(t) + \gamma^2\sum_{i=1}^{r}\sum_{j=1}^{r}\sum_{m=1}^{r}\sum_{n=1}^{r}\mu_i\mu_j\mu_m\mu_n[\tilde{w}^T(t)\tilde{w}(t)]$$

to and from (A.55), we get

$$\tilde{\Delta}V(x(t),\imath) = -\aleph^2(\imath)z^T(t)z(t) + \gamma^2\sum_{i=1}^{r}\sum_{j=1}^{r}\sum_{m=1}^{r}\sum_{n=1}^{r}\mu_i\mu_j\mu_m\mu_n[\tilde{w}^T(t)\tilde{w}(t)]$$
$$+\aleph^2(\imath)z^T(t)z(t) + \sum_{i=1}^{r}\sum_{j=1}^{r}\sum_{m=1}^{r}\sum_{n=1}^{r}\mu_i\mu_j\mu_m\mu_n\begin{bmatrix}\breve{x}(t)\\ \tilde{w}(t)\end{bmatrix}^T \times$$
$$\left(\begin{matrix}\begin{pmatrix}(A_{cl}^{ij}(\imath))^T P(\imath) + P(\imath)A_{cl}^{ij}(\imath)\\ +\sum_{k=1}^{s}\lambda_{\imath k}P(k)\end{pmatrix} & (*)^T\\ (B_{cl}^{ij}(\imath))^T P(\imath) & -\gamma^2 I\end{matrix}\right)\begin{bmatrix}\breve{x}(t)\\ \tilde{w}(t)\end{bmatrix}. \quad (A.56)$$

Now let us consider the following terms:

$$\gamma^2\sum_{i=1}^{r}\sum_{j=1}^{r}\sum_{m=1}^{r}\sum_{n=1}^{r}\mu_i\mu_j\mu_m\mu_n[\tilde{w}^T(t)\tilde{w}(t)] = \gamma^2\sum_{i=1}^{r}\sum_{j=1}^{r}\sum_{m=1}^{r}\sum_{n=1}^{r}\mu_i\mu_j\mu_m\mu_n \times$$
$$\begin{bmatrix}\frac{1}{\delta(\imath)}F(x(t),\imath,t)H_{1_i}(\imath)x(t)\\ F(x(t),\imath,t)H_{2_i}(\imath)w(t)\\ \frac{1}{\delta(\imath)}F(x(t),\imath,t)H_{3_i}(\imath)\hat{C}_j(\imath)\hat{x}(t)\\ \frac{1}{\delta(\imath)}F(x(t),\imath,t)H_{5_i}(\imath)x(t)\\ w(t)\\ F(x(t),\imath,t)H_{7_i}(\imath)w(t)\end{bmatrix}^T\begin{bmatrix}\frac{1}{\delta(\imath)}F(x(t),\imath,t)H_{1_m}(\imath)x(t)\\ F(x(t),\imath,t)H_{2_m}(\imath)w(t)\\ \frac{1}{\delta(\imath)}F(x(t),\imath,t)H_{3_m}(\imath)\hat{C}_n(\imath)\hat{x}(t)\\ \frac{1}{\delta(\imath)}F(x(t),\imath,t)H_{5_m}(\imath)x(t)\\ w(t)\\ F(x(t),\imath,t)H_{7_m}(\imath)w(t)\end{bmatrix}$$
$$\leq \frac{\gamma^2\rho^2(\imath)}{\delta^2(\imath)}\sum_{i=1}^{r}\sum_{j=1}^{r}\sum_{m=1}^{r}\sum_{n=1}^{r}\mu_i\mu_j\mu_m\mu_n\breve{x}^T(t) \times$$
$$\left[\begin{pmatrix}H_{1_i}^T(\imath)H_{1_m}(\imath)\\ +H_{5_i}^T(\imath)H_{5_m}(\imath)\end{pmatrix}\begin{pmatrix}\hat{C}_j^T(\imath)H_{3_i}^T(\imath)\times\\ H_{3_m}(\imath)\hat{C}_n(\imath)\end{pmatrix}\right]\breve{x}(t) + \aleph^2(\imath)\gamma^2 w^T(t)w(t)$$
$$(A.57)$$

and

$$\aleph^2(\imath)z^T(t)z(t) = \aleph^2(\imath)\sum_{i=1}^{r}\sum_{j=1}^{r}\sum_{m=1}^{r}\sum_{n=1}^{r}\mu_i\mu_j\mu_m\mu_n\breve{x}^T(t)\times$$
$$\left[C_{1_i}(\imath) + F(x(t),\imath,t)H_{4_i}(\imath) \quad D_{12_i}(\imath)\hat{C}_j(\imath) + F(x(t),\imath,t)H_{6_i}(\imath)\hat{C}_j(\imath)\right]^T \times$$
$$\left[C_{1_m}(\imath) + F(x(t),\imath,t)H_{4_m}(\imath) \quad D_{12_m}(\imath)\hat{C}_n(\imath) + F(x(t),\imath,t)H_{6_m}(\imath)\hat{C}_n(\imath)\right]\breve{x}(t)$$
$$\leq \sum_{i=1}^{r}\sum_{j=1}^{r}\sum_{m=1}^{r}\sum_{n=1}^{r}\mu_i\mu_j\mu_m\mu_n\Big(2\aleph^2(\imath)\breve{x}^T(t)\left[C_{1_i}(\imath) \quad D_{12_i}(\imath)\hat{C}_j(\imath)\right]^T \times$$
$$\left[C_{1_m}(\imath) \quad D_{12_m}(\imath)\hat{C}_n(\imath)\right]\breve{x}(t) + 2\aleph^2(\imath)\rho^2(\imath)\tilde{x}^T(t)\times$$
$$\left[H_{4_i}(\imath) \quad H_{6_i}(\imath)\hat{C}_j(\imath)\right]^T\left[H_{4_m}(\imath) \quad H_{6_m}(\imath)\hat{C}_n(\imath)\right]\breve{x}(t)\Big) \quad (A.58)$$

where $\aleph(\imath) \geq \Big(1 + \rho^2(\imath)\Big[\|H_{2_i}^T(\imath)H_{2_j}(\imath)\| + \|H_{7_i}^T(\imath)H_{7_j}(\imath)\|\Big]\Big)^{\frac{1}{2}}$. Hence,

$$\gamma^2 \sum_{i=1}^{r}\sum_{j=1}^{r}\sum_{m=1}^{r}\sum_{n=1}^{r} \mu_i\mu_j\mu_m\mu_n[\tilde{w}^T(t)\tilde{w}(t)] + \aleph^2(\imath)z^T(t)z(t)$$

$$\leq \sum_{i=1}^{r}\sum_{j=1}^{r}\sum_{m=1}^{r}\sum_{n=1}^{r} \mu_i\mu_j\mu_m\mu_n\Big(\breve{x}^T(t)\Big[\tilde{C}_{1_i}(\imath) \quad \tilde{D}_{12_i}(\imath)\hat{C}_j(\imath)\Big]^T \times$$

$$\Big[\tilde{C}_{1_m}(\imath) \quad \tilde{D}_{12_m}(\imath)\hat{C}_n(\imath)\Big]\breve{x}(t)\Big) + \aleph^2(\imath)\gamma^2 w^T(t)w(t) \tag{A.59}$$

where

$$\tilde{C}_{1_i}(\imath) = \Big[\tfrac{\gamma\rho(\imath)}{\delta(\imath)}H_{1_i}^T(\imath) \;\; 0 \;\; \tfrac{\gamma\rho(\imath)}{\delta(\imath)}H_{5_i}^T(\imath) \;\; \sqrt{2}\aleph(\imath)\rho(\imath)H_{4_i}^T(\imath) \;\; \sqrt{2}\aleph(\imath)C_{1_i}^T(\imath)\Big]^T$$

$$\tilde{D}_{12_i}(\imath) = \Big[0 \;\; \tfrac{\gamma\rho(\imath)}{\delta(\imath)}H_{3_i}^T(\imath) \;\; 0 \;\; \sqrt{2}\aleph(\imath)\rho(\imath)H_{6_i}^T(\imath) \;\; \sqrt{2}\aleph(\imath)D_{12_i}^T(\imath)\Big]^T.$$

Substituting (A.59) into (A.56), we have

$$\tilde{\Delta}V(x(t),\imath) \leq -\aleph^2(\imath)z^T(t)z(t) + \gamma^2\aleph^2(\imath)w^T(t)w(t) + \sum_{i=1}^{r}\sum_{j=1}^{r}\sum_{m=1}^{r}\sum_{n=1}^{r} \mu_i\mu_j\mu_m\mu_n \begin{bmatrix} x(t) \\ \tilde{w}(t) \end{bmatrix}^T \Omega_{ijmn}(\imath) \begin{bmatrix} x(t) \\ \tilde{w}(t) \end{bmatrix} \tag{A.60}$$

where

$$\Omega_{ijmn}(\imath) = \begin{pmatrix} \begin{pmatrix} (A_{cl}^{ij}(\imath))^T P(\imath) + P(\imath)A_{cl}^{ij}(\imath) \\ +(C_{cl}^{ij}(\imath))^T C_{cl}^{mn}(\imath) + \sum_{k=1}^{s}\lambda_{\imath k}P(k) \end{pmatrix} & (*)^T \\ (B_{cl}^{ij}(\imath))^T P(\imath) & -\gamma^2 I \end{pmatrix}. \tag{A.61}$$

Using the fact

$$\sum_{i=1}^{r}\sum_{j=1}^{r}\sum_{m=1}^{r}\sum_{n=1}^{r} \mu_i\mu_j\mu_m\mu_n M_{ij}^T(\imath)N_{mn}(\imath) \leq \frac{1}{2}\sum_{i=1}^{r}\sum_{j=1}^{r}\mu_i\mu_j[M_{ij}^T(\imath)M_{ij}(\imath) + N_{ij}(\imath)N_{ij}^T(\imath)],$$

we can rewrite (A.61) as follows:

$$\tilde{\Delta}V(x(t),\imath) \leq -\aleph^2(\imath)z^T(t)z(t) + \gamma^2\aleph^2(\imath)w^T(t)w(t) + \sum_{i=1}^{r}\sum_{j=1}^{r}\mu_i\mu_j \begin{bmatrix} x(t) \\ \tilde{w}(t) \end{bmatrix}^T \Omega_{ij}(\imath) \begin{bmatrix} x(t) \\ \tilde{w}(t) \end{bmatrix} \tag{A.62}$$

where

$$\Omega_{ij}(\imath) = \begin{pmatrix} \begin{pmatrix} (A_{cl}^{ij}(\imath))^T P(\imath) + P(\imath)A_{cl}^{ij}(\imath) \\ +(C_{cl}^{ij}(\imath))^T C_{cl}^{ij}(\imath) + \sum_{k=1}^{s}\lambda_{\imath k}P(k) \end{pmatrix} & (*)^T \\ (B_{cl}^{ij}(\imath))^T P(\imath) & -\gamma^2 I \end{pmatrix}. \tag{A.63}$$

Note that (A.63) is the Schur complement of (5.26). Using the inequality (5.26), we have

$$\tilde{\Delta}V(x(t),\imath) < -\aleph^2(\imath)z^T(t)z(t) + \gamma^2\aleph^2(\imath)w^T(t)w(t). \qquad (A.64)$$

Applying the operator $\mathbf{E}[\int_0^{T_f}(\cdot)dt]$ on both sides of (A.64), we obtain

$$\mathbf{E}\left[\int_0^{T_f}\tilde{\Delta}V(x(t),\imath)dt\right] < \mathbf{E}\left[\int_0^{T_f}(-\aleph^2(\imath)z^T(t)z(t) + \gamma^2\aleph^2(\imath)w^T(t)w(t))dt\right]. \qquad (A.65)$$

From the Dynkin's formula [75], it follows that

$$\mathbf{E}\left[\int_0^{T_f}\tilde{\Delta}V(x(t),\imath)dt\right] = \mathbf{E}[V(x(T_f),\imath(T_f))] - \mathbf{E}[V(x(0),\imath(0))]. \qquad (A.66)$$

Substitute (A.66) into (A.65) yields

$$0 < \mathbf{E}\left[\int_0^{T_f}(-\aleph^2(\imath)z^T(t)z(t) + \gamma^2\aleph^2(\imath)w^T(t)w(t))dt\right] \\ -\mathbf{E}[V(x(T_f),\imath(T_f))] + \mathbf{E}[V(x(0),\imath(0))].$$

Using (A.49) and the fact that $V(x(0) = 0, \imath(0)) = 0$ and $V(x(T_f),\imath(T_f)) > 0$, we have

$$\mathbf{E}\left[\int_0^{T_f}\left\{z^T(t)z(t) - \gamma^2 w^T(t)w(t)\right\}dt\right] < 0. \qquad (A.67)$$

Hence the inequality (5.5) holds. This completes the proof of Lemma 3. ■

Proof of Theorem 8.2

Suppose the inequalities (8.36)-(8.38) hold, then the matrices X_0 and Y_0 are of the following forms:

$$X_0 = \begin{pmatrix} X_1 & X_2 \\ 0 & X_3 \end{pmatrix} \text{ and } Y_0 = \begin{pmatrix} Y_1 & Y_2 \\ 0 & Y_3 \end{pmatrix}$$

with $X_1 = X_1^T > 0$, $X_3 = X_3^T > 0$, $Y_1 = Y_1^T > 0$ and $Y_3 = Y_3^T > 0$. Substituting X_0 and Y_0 into (8.47), respectively, we have

$$X_\varepsilon = \left\{X_0 + \varepsilon\tilde{X}\right\}E_\varepsilon = \begin{pmatrix} X_1 & \varepsilon X_2 \\ \varepsilon X_2^T & \varepsilon X_3 \end{pmatrix}. \qquad (A.68)$$

and

$$Y_\varepsilon^{-1} = \left\{Y_0^{-1} + \varepsilon N_\varepsilon\right\}E_\varepsilon = \begin{pmatrix} Y_1^{-1} & -\varepsilon Y^{-1}Y_2Y_3^{-1} \\ -\varepsilon(Y^{-1}Y_2Y_3^{-1})^T & \varepsilon Y_3^{-1} \end{pmatrix}. \qquad (A.69)$$

Clearly, $X_\varepsilon = X_\varepsilon^T$, and $Y_\varepsilon^{-1} = (Y_\varepsilon^{-1})^T$. Knowing the fact that the inverse of a symmetric matrix is a symmetric matrix, we learn that Y_ε is a symmetric matrix. Using the matrix inversion lemma, we can see that

$$Y_\varepsilon = E_\varepsilon^{-1}\left\{Y_0 + \varepsilon\tilde{Y}\right\} \tag{A.70}$$

where $\tilde{Y} = Y_0 N_\varepsilon (I + \varepsilon Y_0 N_\varepsilon)^{-1} Y_0$. Employing the Schur complement, one can show that there exists a sufficiently small $\hat{\varepsilon}$ such that for $\varepsilon \in (0, \hat{\varepsilon}]$, (8.26) and (8.27) hold.

Now, we need to show that

$$\begin{pmatrix} X_\varepsilon & I \\ I & Y_\varepsilon \end{pmatrix} > 0. \tag{A.71}$$

By the Schur complement, it is equivalent to showing that

$$X_\varepsilon - Y_\varepsilon^{-1} > 0. \tag{A.72}$$

Substituting (A.68) and (A.69) into the left hand side of (A.72), we get

$$\begin{bmatrix} X_1 - Y_1^{-1} & \varepsilon(X_2 + Y_1^{-1}Y_2Y_3^{-1}) \\ \varepsilon(X_2 + Y_1^{-1}Y_2Y_3^{-1})^T & \varepsilon(X_3 - Y_3^{-1}) \end{bmatrix}. \tag{A.73}$$

The Schur complement of (8.36) is

$$\begin{bmatrix} X_1 - Y_1^{-1} & 0 \\ 0 & X_3 - Y_3^{-1} \end{bmatrix} > 0. \tag{A.74}$$

According to (A.74), we learn that

$$X_1 - Y_1^{-1} > 0 \qquad \text{and} \qquad X_3 - Y_3^{-1} > 0. \tag{A.75}$$

Using (A.75) and the Schur complement, it can be shown that there exists a sufficiently small $\hat{\varepsilon} > 0$ such that for $\varepsilon \in (0, \hat{\varepsilon}]$, (8.25) holds.

Next, employing (A.68), (A.69) and (A.70), the controller's matrices given in (8.34) can be re-expressed as follows:

$$\begin{aligned} \mathcal{B}_i(\varepsilon) &= \left[Y_0^{-1} - X_0\right]\hat{B}_i + \varepsilon\left[N_\varepsilon - \tilde{X}\right]\hat{B}_i \triangleq \mathcal{B}_{0_i} + \varepsilon\mathcal{B}_{\varepsilon_i} \\ \mathcal{C}_i(\varepsilon) &= \hat{C}_i Y_0^T + \varepsilon\hat{C}_i\tilde{Y}^T \triangleq \mathcal{C}_{0_i} + \varepsilon\mathcal{C}_{\varepsilon_i}. \end{aligned} \tag{A.76}$$

Substituting (A.68), (A.69), (A.70) and (A.76) into (8.32) and (8.33), and pre-post multiplying (8.32) by $\begin{pmatrix} E_\varepsilon & 0 \\ 0 & I \end{pmatrix}$, we, respectively, obtain

$$\Psi_{11_{ij}} + \psi_{11_{ij}} \text{ and } \Psi_{22_{ij}} + \psi_{22_{ij}} \tag{A.77}$$

where the ε-independent linear matrices $\Psi_{11_{ij}}$ and $\Psi_{22_{ij}}$ are defined in (8.43) and (8.44), respectively and the ε-dependent linear matrices are

$$\psi_{11_{ij}} = \varepsilon \begin{pmatrix} A_i\tilde{Y}^T + \tilde{Y}A_i^T + B_{2_i}\mathcal{C}_{\varepsilon_j} + \mathcal{C}_{\varepsilon_i}^T B_{2_j}^T & (*)^T \\ \left[\tilde{Y}\tilde{C}_{1_i}^T + \mathcal{C}_{\varepsilon_i}^T \tilde{D}_{12_j}^T\right]^T & 0 \end{pmatrix} \tag{A.78}$$

$$\psi_{22_{ij}} = \varepsilon \begin{pmatrix} A_i^T\tilde{X} + \tilde{X}^T A_i + \mathcal{B}_{\varepsilon_i} C_{2_j} + C_{2_i}^T \mathcal{B}_{\varepsilon_j}^T & (*)^T \\ \left[\tilde{X}\tilde{B}_{1_i} + \mathcal{B}_{\varepsilon_i}\tilde{D}_{21_j}\right]^T & 0 \end{pmatrix}. \tag{A.79}$$

Note that the ε-dependent linear matrices tend to zero when ε approaches zero.

Employing (8.39)–(8.42) and knowing the fact that for any given negative definite matrix $\mathcal{W}$, there exists an $\varepsilon > 0$ such that $\mathcal{W} + \varepsilon I < 0$, one can show that there exists a sufficiently small $\hat{\varepsilon} > 0$ such that for $\varepsilon \in (0, \hat{\varepsilon}]$, (8.28)–(8.31) hold. Since (8.25)-(8.31) hold, using Lemma 8.2, the inequality (3.3) holds. ■

Proof of Theorem 10.2

Suppose the inequalities (10.56)-(10.58) hold, then the matrices $X_0(\imath)$ and $Y_0(\imath)$ are of the following forms:

$$X_0(\imath) = \begin{pmatrix} X_1(\imath) & X_2(\imath) \\ 0 & X_3(\imath) \end{pmatrix} \text{ and } Y_0(\imath) = \begin{pmatrix} Y_1(\imath) & Y_2(\imath) \\ 0 & Y_3(\imath) \end{pmatrix}$$

with $X_1(\imath) = X_1^T(\imath) > 0$, $X_3(\imath) = X_3^T(\imath) > 0$, $Y_1(\imath) = Y_1^T(\imath) > 0$ and $Y_3(\imath) = Y_3^T(\imath) > 0$. Substituting $X_0(\imath)$ and $Y_0(\imath)$ into (10.67)-(10.68), respectively, we have

$$X_\varepsilon(\imath) = \begin{pmatrix} X_1(\imath) & \varepsilon X_2(\imath) \\ \varepsilon X_2^T(\imath) & \varepsilon X_3(\imath) \end{pmatrix} \tag{A.80}$$

and

$$Y_\varepsilon^{-1}(\imath) = \begin{pmatrix} Y_1^{-1}(\imath) & -\varepsilon Y^{-1}(\imath)Y_2(\imath)Y_3^{-1}(\imath) \\ -\varepsilon\left(Y^{-1}(\imath)Y_2(\imath)Y_3^{-1}(\imath)\right)^T & \varepsilon Y_3^{-1}(\imath) \end{pmatrix}. \tag{A.81}$$

Clearly, $X_\varepsilon(\imath) = X_\varepsilon^T(\imath)$, and $Y_\varepsilon^{-1}(\imath) = (Y_\varepsilon^{-1}(\imath))^T$. Knowing the fact that the inverse of a symmetric matrix is a symmetric matrix, we learn that $Y_\varepsilon(\imath)$ is a symmetric matrix. Using the matrix inversion lemma, we can see that $Y_\varepsilon(\imath) = E_\varepsilon^{-1}\left\{Y_0(\imath) + \varepsilon\tilde{Y}(\imath)\right\}$ where $\tilde{Y}(\imath) = Y_0(\imath)N_\varepsilon(\imath)(I + \varepsilon Y_0(\imath)N_\varepsilon(\imath))^{-1}Y_0(\imath)$. Employing the Schur complement, one can show that there exists a sufficiently small $\hat{\varepsilon}$ such that for $\varepsilon \in (0, \hat{\varepsilon}]$, (10.46) and (10.47) hold.

Now, we need to show that

$$\begin{pmatrix} X_\varepsilon(\imath) & I \\ I & Y_\varepsilon(\imath) \end{pmatrix} > 0. \tag{A.82}$$

By the Schur complement, it is equivalent to showing that

$$X_\varepsilon(\imath) - Y_\varepsilon^{-1}(\imath) > 0. \tag{A.83}$$

Substituting (A.80) and (A.81) into the left hand side of (A.83), we get

$$\begin{bmatrix} X_1(\imath) - Y_1^{-1}(\imath) & \varepsilon(X_2(\imath) + Y_1^{-1}(\imath)Y_2(\imath)Y_3^{-1}(\imath)) \\ \varepsilon(X_2(\imath) + Y_1^{-1}(\imath)Y_2(\imath)Y_3^{-1}(\imath))^T & \varepsilon(X_3(\imath) - Y_3^{-1}(\imath)) \end{bmatrix}. \tag{A.84}$$

The Schur complement of (10.56) is

$$\begin{bmatrix} X_1(\imath) - Y_1^{-1}(\imath) & 0 \\ 0 & X_3(\imath) - Y_3^{-1}(\imath) \end{bmatrix} > 0. \tag{A.85}$$

According to (A.85), we learn that

$$X_1(\imath) - Y_1^{-1}(\imath) > 0 \quad \text{and} \quad X_3(\imath) - Y_3^{-1}(\imath) > 0. \tag{A.86}$$

Using (A.86) and the Schur complement, it can be shown that there exists a sufficiently small $\hat{\varepsilon} > 0$ such that for $\varepsilon \in (0, \hat{\varepsilon}]$, (10.45) holds.

Next, employing (A.80) and (A.81), the controller's matrices given in (10.54) can be re-expressed as follows:

$$\begin{aligned} \mathcal{B}_i(\imath, \varepsilon) &= \left[Y_0^{-1}(\imath) - X_0(\imath)\right]\hat{B}_i(\imath) + \varepsilon\left[N_\varepsilon(\imath) - \tilde{X}(\imath)\right]\hat{B}_i(\imath) \triangleq \mathcal{B}_{0_i}(\imath) + \varepsilon\mathcal{B}_{\varepsilon_i}(\imath) \\ \mathcal{C}_i(\imath, \varepsilon) &= \hat{C}_i(\imath)Y_0^T(\imath) + \varepsilon\hat{C}_i(\imath)\tilde{Y}^T(\imath) \triangleq \mathcal{C}_{0_i}(\imath) + \varepsilon\mathcal{C}_{\varepsilon_i}(\imath). \end{aligned} \tag{A.87}$$

Substituting (A.80),(A.81) and (A.87) into (10.52) and (10.53), and pre-post multiplying (10.52) by $\begin{pmatrix} E_\varepsilon & 0 & 0 \\ 0 & I & 0 \\ 0 & 0 & I \end{pmatrix}$, we respectively, obtain

$$\Psi_{11_{ij}}(\imath) + \psi_{11_{ij}}(\imath) \text{ and } \Psi_{22_{ij}}(\imath) + \psi_{22_{ij}}(\imath) \tag{A.88}$$

where the ε-independent linear matrices $\Psi_{11_{ij}}(\imath)$ and $\Psi_{22_{ij}}(\imath)$ are defined in (10.63) and (10.64), respectively, and the ε-dependent linear matrices are

$$\psi_{11_{ij}}(\imath) = \varepsilon \begin{pmatrix} \begin{pmatrix} A_i(\imath)\tilde{Y}^T(\imath) + \tilde{Y}(\imath)A_i^T(\imath) \\ +B_{2_i}(\imath)\mathcal{C}_{\varepsilon_j}(\imath) + \mathcal{C}_{\varepsilon_i}^T(\imath)B_{2_j}^T(\imath) \\ +\lambda_{\imath\imath}\hat{\tilde{Y}}(\imath) \end{pmatrix} & (*)^T & (*)^T \\ \tilde{C}_{1_i}(\imath)\tilde{Y}^T(\imath) + \tilde{D}_{12_i}(\imath)\mathcal{C}_{\varepsilon_j}(\imath) & 0 & (*)^T \\ \tilde{\mathcal{J}}^T(\imath) & 0 & -\tilde{\mathcal{Y}}(\imath) \end{pmatrix} \tag{A.89}$$

and

$$\psi_{22_{ij}}(\imath) = \varepsilon \begin{pmatrix} \begin{pmatrix} A_i^T(\imath)\tilde{X}^T(\imath) + \tilde{X}(\imath)A_i(\imath) \\ +\mathcal{B}_{\varepsilon_i}(\imath)C_{2_j}(\imath) + C_{2_i}^T(\imath)\mathcal{B}_{\varepsilon_j}^T(\imath) \\ +\sum_{k=1}^{s}\lambda_{\imath k}\hat{\tilde{X}}(k) \end{pmatrix} & (*)^T \\ \tilde{B}_{1_i}^T(\imath)\tilde{X}^T(\imath) + \tilde{D}_{21_i}^T(\imath)\mathcal{B}_{\varepsilon_j}^T(\imath) & 0 \end{pmatrix} \tag{A.90}$$

where $\tilde{\mathcal{J}}(\imath) = \left[\sqrt{\lambda_{1\imath}}\hat{\tilde{Y}}(\imath) \quad \cdots \quad \sqrt{\lambda_{(i-1)\imath}}\hat{\tilde{Y}}(\imath) \quad \sqrt{\lambda_{(i+1)\imath}}\hat{\tilde{Y}}(\imath) \quad \cdots \quad \sqrt{\lambda_{s\imath}}\hat{\tilde{Y}}(\imath)\right]$, $\tilde{\mathcal{Y}}(\imath) = \text{diag}\left\{\hat{\tilde{Y}}(1), \quad \cdots, \quad \hat{\tilde{Y}}(\imath-1), \quad \hat{\tilde{Y}}(\imath+1), \quad \cdots \quad, \hat{\tilde{Y}}(s)\right\}$, $\hat{\tilde{X}}(k) = \frac{\tilde{X}(k)+\tilde{X}^T(k)}{2}$ and $\hat{\tilde{Y}}(\imath) = \frac{\tilde{Y}(\imath)+\tilde{Y}^T(\imath)}{2}$. Note that the ε-dependent linear matrices tend to zero when ε approaches zero.

Employing (10.59)-(10.62) and knowing the fact that for any given negative definite matrix $\mathcal{W}$, there exists an $\varepsilon > 0$ such that $\mathcal{W} + \varepsilon I < 0$, one can show that there exists a sufficiently small $\hat{\varepsilon} > 0$ such that for $\varepsilon \in (0, \hat{\varepsilon}]$, (10.48)-(10.51) hold. Since (10.45)-(10.51) hold, using Lemma 9, the inequality (5.5) holds. ■

References

1. J. A. Ball and J. W. Helton, "$\mathcal{H}_\infty$ control for nonlinear plants: Connection with differential games," in *Proc. IEEE Conf. Decision and Contr.*, pp. 956–962, 1989.
2. J. A. Ball, J. W. Helton, and M. L. Walker, "$\mathcal{H}_\infty$ control for nonlinear systems with output feedback," *IEEE Trans. Automat. Contr.*, vol. 38, pp. 546–559, 1993.
3. N. Berman and U. Shaked, "$\mathcal{H}_\infty$ nonlinear filtering," *Int. J. Robust and Nonlinear Contr.*, vol. 6, pp. 281–296, 1996.
4. T. Basar and G. J. Olsder, *Dynamic Noncooperative Game Theory.* New York: Academic Press, 1982.
5. A. J. van der Schaft, "$\mathcal{L}_2$-gain analysis of nonlinear systems and nonlinear state feedback $\mathcal{H}_\infty$ control," *IEEE Trans. Automat. Contr.*, vol. 37, pp. 770–784, 1992.
6. A. Isidori, "Feedback control of nonlinear systems," in *Proc. First European Contr. Conf.*, pp. 1001–1012, 1991.
7. A. Isidori and A. Astolfi, "Disturbance attenuation and $\mathcal{H}_\infty$- control via measurement feedback in nonlinear systems," *IEEE Trans. Automat. Contr.*, vol. 37, pp. 1283–1293, 1992.
8. J. L. Willems, "The circle criterion and quadratic Lyapunov functions for stability analysis," *IEEE Trans. Automat. Contr.*, vol. 18, pp. 184–186, 1973.
9. J. L. Willems, "On the existence of nonpositive solution to the Riccati equation," *IEEE Trans. Automat. Contr.*, vol. 19, pp. 592–593, 1974.
10. V. A. Yakubovich, "The solution of certain matrix inequalities in automatic control theory," *Soviet Math. Dokl.*, vol. 3, pp. 620–623, 1962.
11. V. A. Yakubovich, "Solution of certain matrix inequalities encountered in nonlinear control theory," *Soviet Math. Dokl.*, vol. 5, pp. 652–656, 1964.
12. V. A. Yakubovich, "The method of matrix inequalities in the stability theory of nonlinear control system, I," *Automation and Remote Contr.*, vol. 25, pp. 905–917, 1967.
13. V. A. Yakubovich, "The method of matrix inequalities in the stability theory of nonlinear control system, II," *Automation and Remote Contr.*, vol. 26, pp. 577–592, 1967.
14. V. A. Yakubovich, "The method of matrix inequalities in the stability theory of nonlinear control system, III," *Automation and Remote Contr.*, vol. 26, pp. 753–763, 1967.

15. A. H. Zak and C. A. Maccarley, "State-feedback control of nonlinear systems," *Int. J. Contr.*, vol. 43, pp. 1497–1514, 1986.
16. S. Suzuki, A. Isidori, and T. J. Tarn, "$\mathcal{H}_\infty$ control of nonlinear systems with sampled measurements," *J. Math. Systems, Estimation, and Contr.*, vol. 5, pp. 1–12, 1995.
17. E. S. Pyatnitskii and V. I. Skorodinskii, "Numerical methods of Lyapunov function construction and their application to the absolute stability problem," *Systems & Control Letters*, vol. 2, pp. 130–135, 1982.
18. V. M. Popov, "Absolute stability of nonlinear system of automatic control," *Automation and Remote Contr.*, vol. 22, pp. 857–875, 1962.
19. H. E. Nusse and C. H. Hommes, "Resolution of chaos with application to a modified Samuelson model," *J. of Economic Dyn. Contr.*, vol. 14, pp. 1–19, 1990.
20. J. L. Willems, "Dissipative dynamical systems Part I: General theory," *Arch. Rational Mech. Anal.*, vol. 45, pp. 321–351, 1972.
21. D. J. Hill and P. J. Moylan, "Dissipative dynamical systems: Basic input-output and state properties," *J. Franklin Inst.*, vol. 309, pp. 327–357, 1980.
22. A. J. van der Schaft, "A state-space approach to nonlinear $\mathcal{H}_\infty$ control," *Systems & Control Letters*, vol. 16, pp. 1–8, 1991.
23. S. K. Nguang, "Robust nonlinear $\mathcal{H}_\infty$ output feedback control," *IEEE Trans Automat. Contr.*, vol. 41, pp. 1003–1008, 1996.
24. S. K. Nguang and M. Fu, "Robust nonlinear $\mathcal{H}_\infty$ filtering," *Automatica*, vol. 32, pp. 1195–1199, 1996.
25. S. K. Nguang and P. Shi, "Nonlinear $\mathcal{H}_\infty$ filtering of sampled-data systems," *Automatica*, vol. 36, pp. 303–310, 2000.
26. S. K. Nguang and P. Shi, "On designing of filters for uncertain sampled-data nonlinear systems," *Systems & Control Letters*, vol. 41, pp. 305–316, 2000.
27. S. K. Nguang and P. Shi, "$\mathcal{H}_\infty$ fuzzy output feedback control design for nonlinear systems: An LMI approach," in *Proc. IEEE Conf. on Decision and Contr.*, (Orlando), pp. 2501–2506, 2001.
28. S. K. Nguang and P. Shi, "$\mathcal{H}^\infty$ fuzzy output feedback control design for nonlinear systems: An LMI approach," *IEEE Trans. Fuzzy Syst.*, vol. 11, pp. 331–340, 2003.
29. J. W. Helton and M. R. James, *Extending $\mathcal{H}^\infty$ control to nonlinear systems.* Philadelphia: SIAM Books, 1999.
30. L. A. Zadeh, "Fuzzy set," *Information and Contr.*, vol. 8, pp. 338–353, 1965.
31. L. A. Zadeh, "Outline of a new approach to the analysis of complex systems and decision processes," *IEEE Trans. Syst. Man, Cybern.*, vol. 3, pp. 28–44, 1973.
32. E. H. Mamdani and S. Assilian, "An experiment in linquistic synthesis with a fuzzy logic controller," *Int. J. Man-Machine-Studies.*, vol. 7, pp. 1–13, 1975.
33. T. Takagi and M. Sugeno, "Fuzzy identification of systems and its applications to model and control," *IEEE Trans. Syst., Man, Cybern.*, vol. 15, pp. 116–132, 1985.
34. S. G. Cao, N. W. Ree, and G. Feng, "Quadratic stabilities analysis and design of continuous-time fuzzy control systems," *Int. J. Syst. Sci.*, vol. 27, pp. 193–203, 1996.
35. X. J. Ma, Z. Q. Sun, and Y. Y. He, "Analysis and design of fuzzy controller and fuzzy observer," *IEEE. Trans. Fuzzy Syst.*, vol. 6, pp. 41–51, 1998.

36. S. K. Nguang and P. Shi, "Stabilisation of a class of nonlinear time-delay systems using fuzzy models," in *Proc. IEEE Conf. on Decision and Contr.*, (Sydney, Australia), pp. 4415–4419, 2000.
37. J. M. Zhang, R. H. Li, and P. A. Zhang, "Stability analysis and systematic design of fuzzy control system," *Fuzzy Sets Systs.*, vol. 120, pp. 65–72, 2001.
38. S. K. Nguang and P. Shi, "$\mathcal{H}_\infty$ control of fuzzy systems model using linear output controller," *Systems Analysis Modelling and Simulation*, 2001.
39. K. Tanaka and M. Sugeno, "Stability analysis and design of fuzzy control systems," *Fuzzy Sets Systs.*, vol. 45, pp. 135–156, 1992.
40. K. Tanaka and M. Sugeno, "Stability and stabiliability of fuzzy neural linear control systems," *IEEE Trans. Fuzzy Syst.*, vol. 3, pp. 438–447, 1995.
41. K. Tanaka, T. Ikeda, and H. O. Wang, "Robust stabilization of a class of uncertain nonlinear systems via fuzzy control: Quadratic stabilizability, $\mathcal{H}_\infty$ control theory, and linear matrix inequality," *IEEE Trans. Fuzzy Syst.*, vol. 4, pp. 1–13, 1996.
42. K. Tanaka and H. O. Wang, "Fuzzy regulators and fuzzy observers: A linear matrix inequality approach," in *Proc. IEEE Conf. Decision and Contr.*, (San Diego), pp. 1315–1320, 1997.
43. K. Tanaka, T. Taniguchi, and H. O. Wang, "Fuzzy control based on quadratic performance function - a linear matrix inequality approach," in *Proc. IEEE Conf. Decision and Contr.*, (Tampa), pp. 2914–2919, 1998.
44. T. Taniguchi, K. Tanaka, K. Yamafuji, and H. O. Wang, "Fuzzy descriptor systems: Stability analysis and design via LMIs," in *Proc. Amer. Contr. Conf.*, (San Diego), pp. 1827–1831, 1999.
45. J. Joh, Y. H. Chen, and R. Langari, "On the stability issues of linear Takagi-Sugeno fuzzy models," *IEEE Trans. Fuzzy Syst.*, vol. 6, pp. 402–410, 1998.
46. J. Park, J. Kim, and D. Park, "LMI-based design of stabilizing fuzzy controller for nonlinear system described by Takagi-Sugeno fuzzy model," *Fuzzy Sets Systs.*, vol. 122, pp. 73–82, 2001.
47. M. Sugeno and G. T. Kang, "Structure identification of fuzzy model," *Fuzzy Sets Systs.*, vol. 28, pp. 15–33, 1988.
48. M. Teixeira and S. H. Zak, "Stabilizing controller design for uncertain nonlinear systems using fuzzy models," *IEEE Trans. Fuzzy Syst.*, vol. 7, pp. 133–142, 1999.
49. L. X. Wang, "Design and analysis of fuzzy identifiers of nonlinear dynamic systems," *IEEE Trans. Automat. Contr.*, vol. 40, pp. 11–23, 1995.
50. H. O. Wang, K. Tanaka, and M. F. Griffin, "An approach to fuzzy control of nonlinear systems: Stability and design issues," *IEEE Trans. Fuzzy Syst.*, vol. 4, pp. 14–23, 1996.
51. L. X. Wang, *A course in fuzzy systems and control.* Englewood Cliffs, NJ: Prentice-Hall, Inc., 1997.
52. J. Yoneyama, M. Nishikawa, H. Katayama, and A. Ichikawa, "Output stabilization of Takagi-Sugeno fuzzy system," *Fuzzy Sets Systs.*, vol. 111, pp. 253–266, 2000.
53. S. H. Zak, "Stabilizing fuzzy system models using linear controllers," *IEEE Trans. Fuzzy Syst.*, vol. 7, pp. 236–240, 1999.
54. C. L. Chen, P. C. Chen, and C. K. Chen, "Analysis and design of fuzzy control system," *Fuzzy Sets Systs.*, vol. 57, pp. 125–140, 1993.

55. B. S. Chen, C. S. Tseng, and H. J. Uang, "Mixed $\mathcal{H}_2/\mathcal{H}_\infty$ fuzzy output feedback control design for nonlinear dynamic systems: An LMI approach," *IEEE Trans. Fuzzy Syst.*, vol. 8, pp. 249–265, 2000.
56. Z. X. Han and G. Feng, "State-feedback $\mathcal{H}_\infty$ controllers design for fuzzy dynamic system using LMI technique," in *Proc. Fuzzy-IEEE Conf.*, pp. 538–544, 1998.
57. Z. X. Han, G. Feng, B. L. Walcott, and Y. M. Zhang, "$\mathcal{H}_\infty$ controller design of fuzzy dynamic systems with pole placement constraints," in *Proc. Amer. Contr. Conf.*, pp. 1939–1943, 2000.
58. S. K. Nguang and P. Shi, "Fuzzy $\mathcal{H}_\infty$ output feedback control of nonlinear systems under sampled measurements," in *Proc. IEEE Conf. on Decision and Contr.*, (Orlando), pp. 120–126, 2001.
59. H. J. Lee, J. B. Park, and G. Chen, "Robust fuzzy control of nonlinear system with parametric uncertainties," *IEEE. Trans. Fuzzy Syst.*, vol. 9, pp. 369–379, 2001.
60. S. Boyd, L. E. Ghaoui, E. Feron, and V. Balakrishnan, *Linear Matrix Inequalities in Systems and Control Theory.* Philadelphia: SIAM Books, 1994.
61. M. Chilali and P. Gahinet, "$\mathcal{H}_\infty$ design with pole placement constraints: An LMI approach," *IEEE Trans. Automat. Contr.*, vol. 41, pp. 358–367, 1996.
62. M. Chilali, P. Gahinet, and P. Apkarian, "Robust pole placement in LMI regions," *IEEE Trans. Automat. Contr.*, vol. 44, pp. 2257–2270, 1999.
63. P. Gahinet and P. Apkarian, "An LMI-based parametization of all $\mathcal{H}_\infty$ controllers with applications," in *Proc. IEEE Conf. Decision and Contr.*, (Texas), pp. 656–661, 1993.
64. P. Gahinet, "Explicit controller formulars for LMI-based $\mathcal{H}_\infty$ synthesis," in *Proc. Amer. Contr. Conf.*, (Maryland), pp. 2396–2400, 1994.
65. L. Wang and R. Langari, "Building sugeno-type models using fuzzy discretization and orthogonal parameter estimation techniques," *IEEE Trans. Fuzzy Syst.*, vol. 3, pp. 454–458, 1995.
66. T. Taniguchi, K. Tanaka, H. Ohtake, and H. Wang, "Model construction, rule reduction, and robust compensation for generalized form of takagi-sugeno fuzzy systems," *IEEE Trans. Fuzzy Syst.*, vol. 9, pp. 525–538, 2001.
67. T. I. K. Tanaka and H. O. Wang, "An lmi approach to fuzzy controller designs based on relaxed stability conditions," in *Proc. IEEE Int. Conf. Fuzzy Syst. (FUZZ/IEEE)*, (Barcelona), pp. 171–176, 1997.
68. T. M. G. M. Ksontini, F. Delmotte and A. Kamoun, "Disturbance rejection using takagi-sugeno fuzzy model applied to an interconneted tank system," in *Proc. IEEE Int. Conf. Systems, Man and Cybern.*, pp. 3352–3357, 2003.
69. S. P. Sethi and Q. Zhang, *Hierarchical Decision Making in Stochastica Manufacturing Systems.* Boston: Birkhauser, 1994.
70. M. Mariton, *Jump Linear System in Automatic Control.* New York: Dekker, 1990.
71. E. K. Boukas and A. Haurie, "Manufacturing flow control and preventive maintenance: A stochastic control approach," *IEEE Trans. Automat. Contr.*, vol. 35, pp. 1024–1031, 1990.
72. E. K. Boukas, Q. Zhang, and G. Yin, "Robust production and maintenance planning in stochastic manufacturing systems," *IEEE Trans. Automat. Contr.*, vol. 40, pp. 1098–1102, 1995.

73. E. K. Boukas and P. Shi, "Stochastic stability and guaranteed cost control of discrete-time uncertain systems with Markovian jumping parameters," *Int. J. Robust. Contr.*, vol. 8, pp. 1155–1167, 1998.
74. E. K. Boukas, P. Shi, S. K. Nguang, and R. K. Agarwal, "Robust $\mathcal{H}_\infty$ control of a class of nonlinear systems with Markovian jumping parameters," in *Proc. Amer. Contr. Conf.*, pp. 970–976, 1999.
75. E. B. Dynkin, *Markov Process.* Berlin: Springer-Verlag, 1965.
76. Y. Ji and H. J. Chizeck, "Controllability, stabilizability, and continuous-time Markovian jump linear quadratic control," *IEEE Trans. Automat. Contr.*, vol. 35, pp. 777–788, 1990.
77. P. Shi and E. K. Boukas, "$\mathcal{H}_\infty$ control for Markovian jumping linear systems with parametric uncertainty," *J. Optim. Theory Appl.*, vol. 95, pp. 77–99, 1997.
78. N. N. Krasovskii and E. A. Lidskii, "Analysis design of controller in systems with random attributes–Part 1," *Automat. Remote Contr.*, vol. 22, pp. 1021–1025, 1961.
79. O. L. V. Costa and M. D. Fragoso, "Stability results for discrete-time linear systems with Markovian jumping parameters," *J. Math. Anal. Appl.*, vol. 179, pp. 154–178, 1993.
80. C. E. de Souza and M. D. Fragoso, "$\mathcal{H}_\infty$ control for linear systems with Markovian jumping parameters," *Contr. Theory Adv. Tech.*, vol. 9, pp. 457–466, 1993.
81. M. D. S. Aliyu and E. K. Boukas, "$\mathcal{H}_\infty$ control for Markovian jump nonlinear systems," in *Proc. IEEE Conf. Decision and Contr.*, pp. 766–771, 1998.
82. E. K. Boukas and Z. K. Liu, *Deterministic and Stochastic Time Delay Systems.* Boston: Birkhauser, 2002.
83. D. P. de Farias, J. C. Geromel, J. B. R. do Val, and O. L. V. Costa, "Output feedback control of Markov jump linear systems in continuous-time," *IEEE Trans. Automat. Contr.*, vol. 45, pp. 944–949, 2000.
84. W. M. Wonham, "Random differential equations in control theory," *Probabilistic Methods in App. Math.*, vol. 2, pp. 131–212, 1970.
85. S. K. Nguang and W. Assawinchaichote, "$\mathcal{H}_\infty$ fuzzy output feedback control design for nonlinear systems with pole placement constraints: An LMI approach," *IEEE Trans. Fuzzy Systs.*, (to appear), 2006.
86. B. Anderson and S. Vongpantlerd, *Network Analysis and Synthesis: A Modern System Theory Approach.* New Jersey: Prentice-Hall, Inc., 1973.
87. B. D. O. Anderson and J. B. Moore, *Optimal Control: Linear Quadratic Methods.* New Jersey: Prentice-Hall, Inc., 2nd ed., 1990.
88. M. Fu, C. E. de Souza, and L. Xie, "$\mathcal{H}_\infty$ estimation for uncertain systems," *Int. J. Robust Nonlinear Contr.*, vol. 2, pp. 87–105, 1992.
89. M. J. Grimble, "$\mathcal{H}_\infty$ design of optimal linear filters," in *Proc. Int. Symp. on MTNS, Linear Circuit, Systs. and Signal Processing: Theory and Appl.*, pp. 538–544, 1986.
90. K. M. Nagpal and P. P. Khargonekar, "Filtering and smoothing in an $\mathcal{H}_\infty$ setting," *IEEE Trans Automat. Contr.*, vol. 36, pp. 152–166, 1991.
91. S. K. Nguang and W. Assawinchaichote, "$\mathcal{H}_\infty$ filtering for fuzzy dynamical systems with $\mathcal{D}$ stability," *IEEE Trans. Circuit Syst. I*, vol. 50, pp. 1503–1508, 2003.
92. W. P. B. Jr. and D. D. Sworder, "Continuous-time regulation of a class of econometric models," *IEEE Trans. Sys. Man and Cybern.*, vol. 5, pp. 341–346, 1975.

93. D. J. Limebeer and U. Shaked, "New results in $\mathcal{H}_\infty$ filtering," in *Proc. Int. Symp. on MTNS*, (Kobe, Japan), pp. 317–322, 1991.
94. K. W. Chang, "Singular perturbations of general boundary value problem," *Siam J. Math. Anal.*, vol. 3, pp. 520–526, 1972.
95. A. H. Haddad, "Linear filtering of singularly perturbed systems," *IEEE Trans. Automat. Contr.*, vol. 21, pp. 515–519, 1976.
96. M. Corless and L. Glielmo, "Robust of output feedback for a class of singularly perturbed nonlinear systems," in *Proc. IEEE Conf. Decision and Contr.*, (London, England), pp. 1066–1071, 1991.
97. V. Dragan, "Asymptotic expansions for game-theoretic Riccati equations and stabilization with disturbance attenuation for singularly perturbed system," *System & Control Letters*, vol. 20, pp. 455–463, 1993.
98. E. Fridman, "State-feedback $\mathcal{H}^\infty$ control of nonlinear singularly perturbed systems," *Int. J. Robust Nonlinear Contr.*, vol. 11, pp. 1115–1125, 2001.
99. Z. Gajic and M. Lim, "A new filtering method for linear singularly perturbed systems," *IEEE Trans. Automat. Contr.*, vol. 39, pp. 1925–1955, 1994.
100. K. Khalil and Z. Gajic, "Near-opimal regulators for stochastic linear singularly perturbed systems," *IEEE Trans. Automat. Contr.*, vol. 29, pp. 531–541, 1984.
101. K. Khalil, "Output feedback control of linear two-time-scale systems," *IEEE Trans. Automat. Contr.*, vol. 32, pp. 784–792, 1987.
102. P. V. Kokotovic and P. Sannuti, "Singular perturbation methods for reducing the model order in optimal control design," *IEEE Trans. Automat. Contr.*, vol. 13, pp. 377–384, 1968.
103. P. V. Kokotovic, R. E. O'Malley, and P. Sannuti, "Singular perturbations and order reduction in control theory: An overview," *Automatica*, vol. 12, pp. 123–132, 1976.
104. P. V. Kokotovic, H. K. Khalil, and J. O'Reilly, *Singular Perturbation Methods in Control: Analysis and Design.* London: Academic Press, 1986.
105. H. Mukaidani and H. Xu, "Robust $\mathcal{H}_\infty$ control problem for nonstandard singularly perturbed systems and applications," in *Proc. Amer. Contr. Conf.*, (Arlington), pp. 3920–3925, 2001.
106. Z. Pan and T. Basar, "$\mathcal{H}^\infty$–optimal control for singularly perturbed systems Part I: Perfect state measurements," *Automatica*, vol. 29, pp. 401–423, 1993.
107. Z. Pan and T. Basar, "$\mathcal{H}^\infty$-optimal control for singularly perturbed systems Part II: Imperfect state measurements," *IEEE Trans. Automat. Contr.*, vol. 39, pp. 280–299, 1994.
108. D. B. Price, "Comment on linear filtering of singularly perturbed systems," *IEEE Trans. Automat. Contr.*, vol. 24, pp. 675–677, 1979.
109. X. Shen and L. Deng, "Decomposition solution of $\mathcal{H}_\infty$ filter gain in singularly perturbed systems," *Signal Processing*, vol. 55, pp. 313–320, 1996.
110. P. Shi and V. Dragan, "$\mathcal{H}_\infty$ control for singularly perturbed systems with parametric uncertainties," in *Proc. Amer. Contr. Conf.*, (New Maxico), pp. 2120–2124, 1997.
111. P. Shi and V. Dragan, "Asymptotic $\mathcal{H}_\infty$ control of singularly perturbed system with parametric uncertainties," *IEEE Trans. Automat. Contr.*, vol. 44, pp. 1738–1742, 1999.
112. W. C. Su, Z. Gajic, and X. M. Shen, "The exact slow-fast decomposition of the algebraic Riccati equation of singularly perturbed systems," *IEEE Trans. Automat. Contr.*, vol. 37, pp. 1456–1459, 1992.

113. R. Bouyekhf, A. E. Hami, and A. E. Moudni, "Optimal control of a particular class of singularly perturbed nonlinear discrete-time systems," *IEEE Trans. Automat. Contr.*, vol. 46, pp. 1097–1101, 2001.
114. T. Grodt and Z. Gajic, "The recursive reduced-order numerical solution of the singularly perturbed matrix differential Riccati equation," *IEEE Trans. Automat. Contr.*, vol. 43, pp. 751–754, 1998.
115. M. T. Lim and Z. Gajic, "Reduced-order $\mathcal{H}_\infty$ optimal filtering for systems with slow and fast modes," *IEEE Trans. Circuits Syst. I*, vol. 47, pp. 250–254, 2000.
116. H. Oloomi and M. E. Sawan, "The observer-based controller design of discrete-time singularly perturbed systems," *IEEE Trans. Automat. Contr.*, vol. 32, pp. 246–248, 1987.
117. J. O'Reilly, "Two time-scale feedback stabilization of linear time-varying singularly perturbed systems," *J. Franklin Inst.*, vol. 30, pp. 465–474, 1979.
118. B. Porter, "Singularly perturbation methods in the design of observers and stabilizing feedback controllers for multivariables linear systems," *Electron. Lett.*, vol. 10, pp. 494–495, 1974.
119. P. Z. H. Shao and M. E. Sawan, "Robust stability of singularly perturbed systems," *Int. J. Contr.*, vol. 58, pp. 1469–1476, 1993.
120. W. Su, "Sliding surface design for singularly perturbed systems," *Int. J. Contr.*, vol. 72, pp. 990–999, 1999.
121. W. Tan, T. Leung, and Q. Tu, "$\mathcal{H}_\infty$ control for singularly perturbed systems," *Automatica*, vol. 34, pp. 255–260, 1998.
122. A. N. Tikhonov, "On the dependence of the solutions of differential equations on a small parameter," *Mat. Sbornik (Moscow)*, vol. 22, pp. 193–204, 1948.
123. R. E. Kalman, "Lyapunov functions for the problem of Lur'e in automatic control," in *Proc. Nat. Acad. Sci.*, vol. 49, pp. 201–205, 1963.
124. A. I. Lur'e, *Some Nonlinear Problem in the Theory of Automatic Control.* London: H. M. Stationary Off., 1957.
125. V. M. Popov, "One problem in the theory of absolute stability of controlled system," *Automation and Remote Contr.*, vol. 25, pp. 1129–1134, 1964.
126. V. R. Saksena, J. O'Reilly, and P. V. Kokotovic, "Singular perturbations and time-scale methods in control theory: Survey," *Automatica*, vol. 20, pp. 273–293, 1984.
127. J. H. Chow and P. V. Kokotovic, "A decomposition of near-optimum regulators for singularly perturbed systems with slow and fast modes," *IEEE Trans. Automat. Contr.*, vol. 21, pp. 701–705, 1976.
128. M. Suzuki and M. Miura, "Stabilizing feedback controller for singularly perturbed linear constant systems," *IEEE Trans. Automat. Contr.*, vol. 21, pp. 123–124, 1976.
129. V. Dragan, P. Shi, and E. K. Boukas, "Control of singularly perturbed systems with Markovian jump parameters: An $\mathcal{H}_\infty$ approach," *Automatica*, vol. 35, pp. 1369–1378, 1999.
130. E. Fridman, "A descriptor system approach to nonlinear singularly perturbed optimal control problem," *Automatica*, vol. 37, pp. 543–549, 2001.
131. E. Fridman, "$\mathcal{H}^\infty$ control of nonlinear singularly perturbed systems and invariant manifolds," in *New Trends in Dynamic Games and Applications, Series: Annals of International Society on Dynamic Games*, (Boston), pp. 25–45, Birkhauser, 1995.
132. E. Fridman, "Effects of small delays on stability of singularly perturbed systems," *Automatica*, vol. 38, pp. 897–902, 2002.

133. E. Fridman, "Stability of singularly perturbed differential-difference systems: An LMI approach," *Dyn. of Continuous, Discrete and Impulsive Systs.*, vol. 9, pp. 201–212, 2002.
134. Z. Pan and T. Basar, "Time-scale separation and robust controller design for uncertain nonlinear singularly perturbed systems under perfect state measurements," *Int. J. Robust Nonlinear Contr.*, vol. 6, pp. 585–608, 1996.
135. H. Tuan and S. Hosoe, "On linear $\mathcal{H}^\infty$ controllers for a class of singularly perturbed systems," *Automatica*, vol. 35, pp. 735–739, 1999.
136. W. Assawinchaichote and S. K. Nguang, "$\mathcal{H}_\infty$ fuzzy control design for nonlinear singularly perturbed systems with pole placement constraints: An LMI approach," *IEEE Trans. Syst., Man, Cybern. B*, vol. 34, pp. 579–588, 2004.
137. W. Assawinchaichote and S. K. Nguang, "$\mathcal{H}_\infty$ filtering for fuzzy singularly perturbed systems with pole placement constraints: An LMI approach," *IEEE Trans. Signal Processing*, vol. 52, pp. 1659–1667, 2004.
138. P. Gahinet, A. Nemirovski, A. J. Laub, and M. Chilali, *LMI Control Toolbox – For Use with MATLAB.* Massachusetts: The MathWorks,Inc., 1995.
139. S. Mehta and J. Chiasson, "Nonlinear control of a series dc motor: Theory and experiment," *IEEE. Trans. Ind. Electron*, vol. 45, pp. 134–141, 1998.

Lecture Notes in Control and Information Sciences

Edited by M. Thoma, M. Morari

Further volumes of this series can be found on our homepage: springer.com

Vol. 347: Assawinchaichote, W.; Nguang, K.S.; Shi P.
Fuzzy Control and Filter Design for Uncertain Fuzzy Systems
188 p. 2006 [3-540-37011-0]

Vol. 346: Tarbouriech, S.; Garcia, G.; Glattfelder, A.H. (Eds.)
Advanced Strategies in Control Systems with Input and Output Constraints
480 p. 2006 [3-540-37009-9]

Vol. 345: Huang, D.-S.; Li, K.; Irwin, G.W. (Eds.)
Intelligent Computing in Signal Processing and Pattern Recognition
1179 p. 2006 [3-540-37257-1]

Vol. 344: Huang, D.-S.; Li, K.; Irwin, G.W. (Eds.)
Intelligent Control and Automation
1121 p. 2006 [3-540-37255-5]

Vol. 341: Commault, C.; Marchand, N. (Eds.)
Positive Systems
448 p. 2006 [3-540-34771-2]

Vol. 340: Diehl, M.; Mombaur, K. (Eds.)
Fast Motions in Biomechanics and Robotics
500 p. 2006 [3-540-36118-9]

Vol. 339: Alamir, M.
Stabilization of Nonlinear Systems Using Receding-horizon Control Schemes
325 p. 2006 [1-84628-470-8]

Vol. 338: Tokarzewski, J.
Finite Zeros in Discrete Time Control Systems
325 p. 2006 [3-540-33464-5]

Vol. 337: Blom, H.; Lygeros, J. (Eds.)
Stochastic Hybrid Systems
395 p. 2006 [3-540-33466-1]

Vol. 336: Pettersen, K.Y.; Gravdahl, J.T.; Nijmeijer, H. (Eds.)
Group Coordination and Cooperative Control
310 p. 2006 [3-540-33468-8]

Vol. 335: Kozłowski, K. (Ed.)
Robot Motion and Control
424 p. 2006 [1-84628-404-X]

Vol. 334: Edwards, C.; Fossas Colet, E.; Fridman, L. (Eds.)
Advances in Variable Structure and Sliding Mode Control
504 p. 2006 [3-540-32800-9]

Vol. 333: Banavar, R.N.; Sankaranarayanan, V.
Switched Finite Time Control of a Class of Underactuated Systems
99 p. 2006 [3-540-32799-1]

Vol. 332: Xu, S.; Lam, J.
Robust Control and Filtering of Singular Systems
234 p. 2006 [3-540-32797-5]

Vol. 331: Antsaklis, P.J.; Tabuada, P. (Eds.)
Networked Embedded Sensing and Control
367 p. 2006 [3-540-32794-0]

Vol. 330: Koumoutsakos, P.; Mezic, I. (Eds.)
Control of Fluid Flow
200 p. 2006 [3-540-25140-5]

Vol. 329: Francis, B.A.; Smith, M.C.; Willems, J.C. (Eds.)
Control of Uncertain Systems: Modelling, Approximation, and Design
429 p. 2006 [3-540-31754-6]

Vol. 328: Loría, A.; Lamnabhi-Lagarrigue, F.; Panteley, E. (Eds.)
Advanced Topics in Control Systems Theory
305 p. 2006 [1-84628-313-2]

Vol. 327: Fournier, J.-D.; Grimm, J.; Leblond, J.; Partington, J.R. (Eds.)
Harmonic Analysis and Rational Approximation
301 p. 2006 [3-540-30922-5]

Vol. 326: Wang, H.-S.; Yung, C.-F.; Chang, F.-R.
H_∞ Control for Nonlinear Descriptor Systems
164 p. 2006 [1-84628-289-6]

Vol. 325: Amato, F.
Robust Control of Linear Systems Subject to Uncertain
Time-Varying Parameters
180 p. 2006 [3-540-23950-2]

Vol. 324: Christofides, P.; El-Farra, N.
Control of Nonlinear and Hybrid Process Systems
446 p. 2005 [3-540-28456-7]

Vol. 323: Bandyopadhyay, B.; Janardhanan, S.
Discrete-time Sliding Mode Control
147 p. 2005 [3-540-28140-1]

Vol. 322: Meurer, T.; Graichen, K.; Gilles, E.D. (Eds.)
Control and Observer Design for Nonlinear Finite and Inąenite Dimensional Systems
422 p. 2005 [3-540-27938-5]

Vol. 321: Dayawansa, W.P.; Lindquist, A.; Zhou, Y. (Eds.)
New Directions and Applications in Control Theory
400 p. 2005 [3-540-23953-7]

Vol. 320: Steffen, T.
Control Reconfiguration of Dynamical Systems
290 p. 2005 [3-540-25730-6]

Vol. 319: Hofbaur, M.W.
Hybrid Estimation of Complex Systems
148 p. 2005 [3-540-25727-6]

Vol. 318: Gershon, E.; Shaked, U.; Yaesh, I.
H_∞ Control and Estimation of State-multiplicative Linear Systems
256 p. 2005 [1-85233-997-7]

Vol. 317: Ma, C.; Wonham, M.
Nonblocking Supervisory Control of State Tree Structures
208 p. 2005 [3-540-25069-7]

Vol. 316: Patel, R.V.; Shadpey, F.
Control of Redundant Robot Manipulators
224 p. 2005 [3-540-25071-9]

Vol. 315: Herbordt, W.
Sound Capture for Human/Machine Interfaces: Practical Aspects of Microphone Array Signal Processing
286 p. 2005 [3-540-23954-5]

Vol. 314: Gil', M.I.
Explicit Stability Conditions for Continuous Systems
193 p. 2005 [3-540-23984-7]

Vol. 313: Li, Z.; Soh, Y.; Wen, C.
Switched and Impulsive Systems
277 p. 2005 [3-540-23952-9]

Vol. 312: Henrion, D.; Garulli, A. (Eds.)
Positive Polynomials in Control
313 p. 2005 [3-540-23948-0]

Vol. 311: Lamnabhi-Lagarrigue, F.; Loría, A.; Panteley, E. (Eds.)
Advanced Topics in Control Systems Theory
294 p. 2005 [1-85233-923-3]

Vol. 310: Janczak, A.
Identification of Nonlinear Systems Using Neural Networks and Polynomial Models
197 p. 2005 [3-540-23185-4]

Vol. 309: Kumar, V.; Leonard, N.; Morse, A.S. (Eds.)
Cooperative Control
301 p. 2005 [3-540-22861-6]

Vol. 308: Tarbouriech, S.; Abdallah, C.T.; Chiasson, J. (Eds.)
Advances in Communication Control Networks
358 p. 2005 [3-540-22819-5]

Vol. 307: Kwon, S.J.; Chung, W.K.
Perturbation Compensator based Robust Tracking Control and State Estimation of Mechanical Systems
158 p. 2004 [3-540-22077-1]

Vol. 306: Bien, Z.Z.; Stefanov, D. (Eds.)
Advances in Rehabilitation
472 p. 2004 [3-540-21986-2]

Vol. 305: Nebylov, A.
Ensuring Control Accuracy
256 p. 2004 [3-540-21876-9]

Vol. 304: Margaris, N.I.
Theory of the Non-linear Analog Phase Locked Loop
303 p. 2004 [3-540-21339-2]

Vol. 303: Mahmoud, M.S.
Resilient Control of Uncertain Dynamical Systems
278 p. 2004 [3-540-21351-1]

Vol. 302: Filatov, N.M.; Unbehauen, H.
Adaptive Dual Control: Theory and Applications
237 p. 2004 [3-540-21373-2]

Vol. 301: de Queiroz, M.; Malisoff, M.; Wolenski, P. (Eds.)
Optimal Control, Stabilization and Nonsmooth Analysis
373 p. 2004 [3-540-21330-9]

Vol. 300: Nakamura, M.; Goto, S.; Kyura, N.; Zhang, T.
Mechatronic Servo System Control
Problems in Industries and their Theoretical Solutions
212 p. 2004 [3-540-21096-2]

Vol. 299: Tarn, T.-J.; Chen, S.-B.; Zhou, C. (Eds.)
Robotic Welding, Intelligence and Automation
214 p. 2004 [3-540-20804-6]

Vol. 298: Choi, Y.; Chung, W.K.
PID Trajectory Tracking Control for Mechanical Systems
127 p. 2004 [3-540-20567-5]

Vol. 297: Damm, T.
Rational Matrix Equations in Stochastic Control
219 p. 2004 [3-540-20516-0]

Vol. 296: Matsuo, T.; Hasegawa, Y.
Realization Theory of Discrete-Time Dynamical Systems
235 p. 2003 [3-540-40675-1]

Vol. 295: Kang, W.; Xiao, M.; Borges, C. (Eds.)
New Trends in Nonlinear Dynamics and Control, and their Applications
365 p. 2003 [3-540-10474-0]

Vol. 294: Benvenuti, L.; De Santis, A.; Farina, L. (Eds.)
Positive Systems: Theory and Applications (POSTA 2003)
414 p. 2003 [3-540-40342-6]